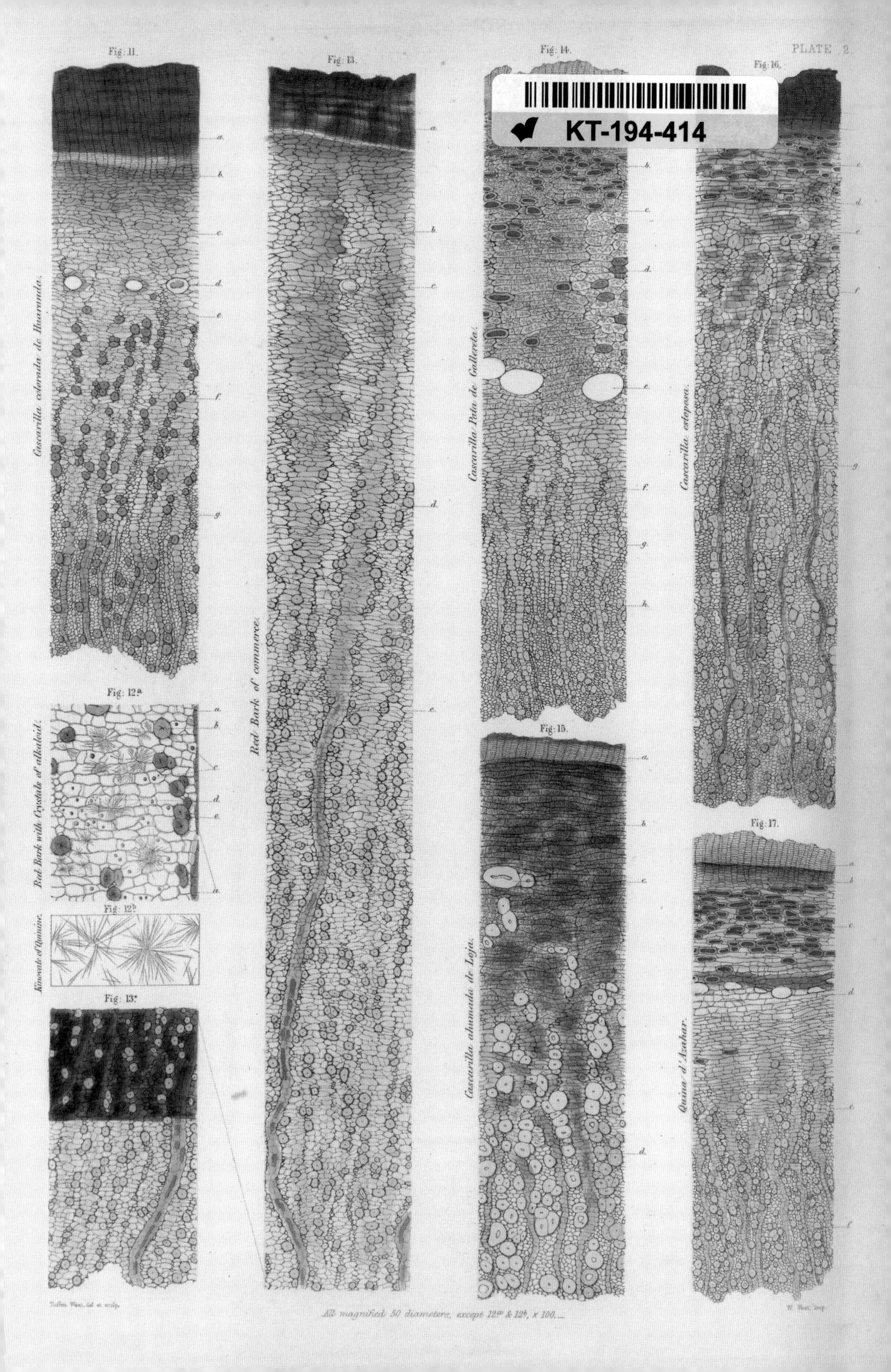

All magnified 50 diameters, except 12a & 12b, × 100.

JUST THE TONIC

KIM WALKER
& MARK NESBITT

JUST THE TONIC

A NATURAL HISTORY *of* TONIC WATER

Kew Publishing
Royal Botanic Gardens, Kew

First published in 2019 by the
Royal Botanic Gardens, Kew, Richmond, Surrey, TW9 3AB, UK
www.kew.org

ISBN 978 1 84246 689 6

Distributed on behalf of the Royal Botanic Gardens, Kew in North America by the University of Chicago Press, 1427 East 60th Street, Chicago, IL 60637, USA.

British Library Cataloguing in Publication Data
A catalogue record for this book is available from the British Library

Design and page layout: Ocky Murray
Production manager: Georgina Hills
Project manager: Lydia White
Copy-editing: Michelle Payne
Proofreading: Matthew Seal

Cover illustration: *Cinchona calisaya*. Allard Pierson, University of Amsterdam [077.580]. End papers: Microscopical observations of cinchona bark and seedlings by Walter Hood Fitch, from J. E. Howard: *Illustrations of the Nueva Quinologia of Pavon*, 1862.

Printed and bound in Italy by L.E.G.O. S.p.A.

For information or to purchase all Kew titles please visit shop.kew.org/kewbooksonline or email publishing@kew.org

Kew's mission is to be the global resource in plant and fungal knowledge and the world's leading botanic garden.

Kew receives approximately one third of its funding from Government through the Department for Environment, Food and Rural Affairs (Defra). All other funding needed to support Kew's vital work comes from members, foundations, donors and commercial activities, including book sales.

CONTENTS

INTRODUCTION

Effervescent in appearance and bitter in flavour, tonic water is today an essential ingredient of the gin and tonic, popular in both urban bars and rural pubs. Yet both the fizz and the taste are reminders of tonic water's medicinal origins. And as so often with ingredients that seem quintessentially British, the origins of tonic water lie far away: on the eastern slopes of the Andes, in the plantations of India and Indonesia and in the spas of Europe. The true story of tonic water is complex and entangled, combining the bitter and antimalarial quinine extracts of the cinchona 'fever' trees native to Latin America, with the popularity of carbonated water as a healthy drink in Europe.

Naturally, such a rich story has attracted many legends and many stories that are too good to be true. In this book we have gone back to original sources, ranging from the rich archives of the Royal Botanic Gardens, Kew, to the freshly digitised newspapers of the former British Empire. The history that emerges places the invention of tonic water alongside many momentous events of the last 500 years: the Spanish invasion of South America and encounter with cinchona trees, one of the few effective treatments for the worldwide scourge of malaria, and the theft of cinchona trees and their establishment in Asian plantations. In parallel there is a lighter narrative: the enthusiasm for alcohol and (later) carbonated water as a means to make quinine palatable, the discovery that gin and tonic is a uniquely refreshing combination, and the rise, fall, and subsequent rise of tonic water in the twentieth century – now inexorably linked to the fortunes of gin.

Both authors carry out research into the uses and history of plants at the Royal Botanic Gardens, Kew. Our starting points are the wonderfully varied specimens held in Kew's Economic Botany Collection and the accompanying botanical library and art collections, perhaps the finest of their kind. Among about 100,000 economic botany specimens, Kew holds about 1,000 bundles of cinchona bark. These were the starting point for our research into the history of quinine, and thus of tonic water. The bundles span 150 years, from the 1780s to the 1930s, a crucial period in which cinchona trees were exposed to Western science and exploration.

Like many others, we see in a glass of tonic water reflections of the wider history of its key ingredient, quinine, and of its most popular admixture, gin. Both quinine and gin are large subjects on which much has been written, and keeping tonic water at the centre has helped us focus what might otherwise be an unwieldy volume. We hope that the next time readers pour a glass, they too will consider the global history contained within each bottle of tonic.

In acknowledgement of quinine's past role as a medicine, several manufacturers of tonic waters and gins donate a proportion of their profits to research into treating malaria, which is still a deadly disease. This is a very practical way of marking a complicated history – a history that does not always do credit to our forebears. We have strived to be honest about this, drawing on recent scholarship so as to avoid an over-romanticised view of the past.

The first part of our book focuses on the story of quinine, the most distinctive

Botanical headpiece from *Dr. Warren's epistle to his friend, of the method and manner of curing the late raging fevers* (1733). Wellcome Collection.

ingredient of tonic water, and malaria, the global disease that it treated so effectively. The second assembles the carbonated water which, mixed with quinine, makes tonic water, alongside its common accompaniments lemon and ice. The third tells the story of gin – the final component required for that iconic drink, the gin and tonic. Although written with an eye for the quirky, we have tried throughout to go back to reliable sources, and to treat serious subjects – for example, the effects of malaria and empire on peoples past and present – seriously. Such a wide-ranging story inevitably depends on the work of many historians, whose publications we cite in *Further Reading*.

In the early twenty-first century tonic water is thriving with dozens of premium brands available, many of which look back to the past for their marketing. It is high time to look in more detail at the fascinating history that comes with every sip of tonic water.

ACKNOWLEDGEMENTS

The authors would like to thank the following: Gina Fullerlove, Georgina Hills, Michelle Payne and Lydia White of Kew Publishing, Ocky Murray for his beautiful design work; Kevin Law-Smith and Jamie He of East Imperial; Toby Sims and Matthew McGivern of London Distillery; Emily Danby for French translation; Jonathan Ray for advice on recipes; Jim Meehan; Kate Teltscher, Helen and William Bynum, Jonathan Drori, Jason Irving, Felix Driver, Alison Foster, Nataly Alassi, Melanie-Jayne Howes and Briony Hudson for commenting on the text. Responsibility for any remaining errors rests with the authors alone.

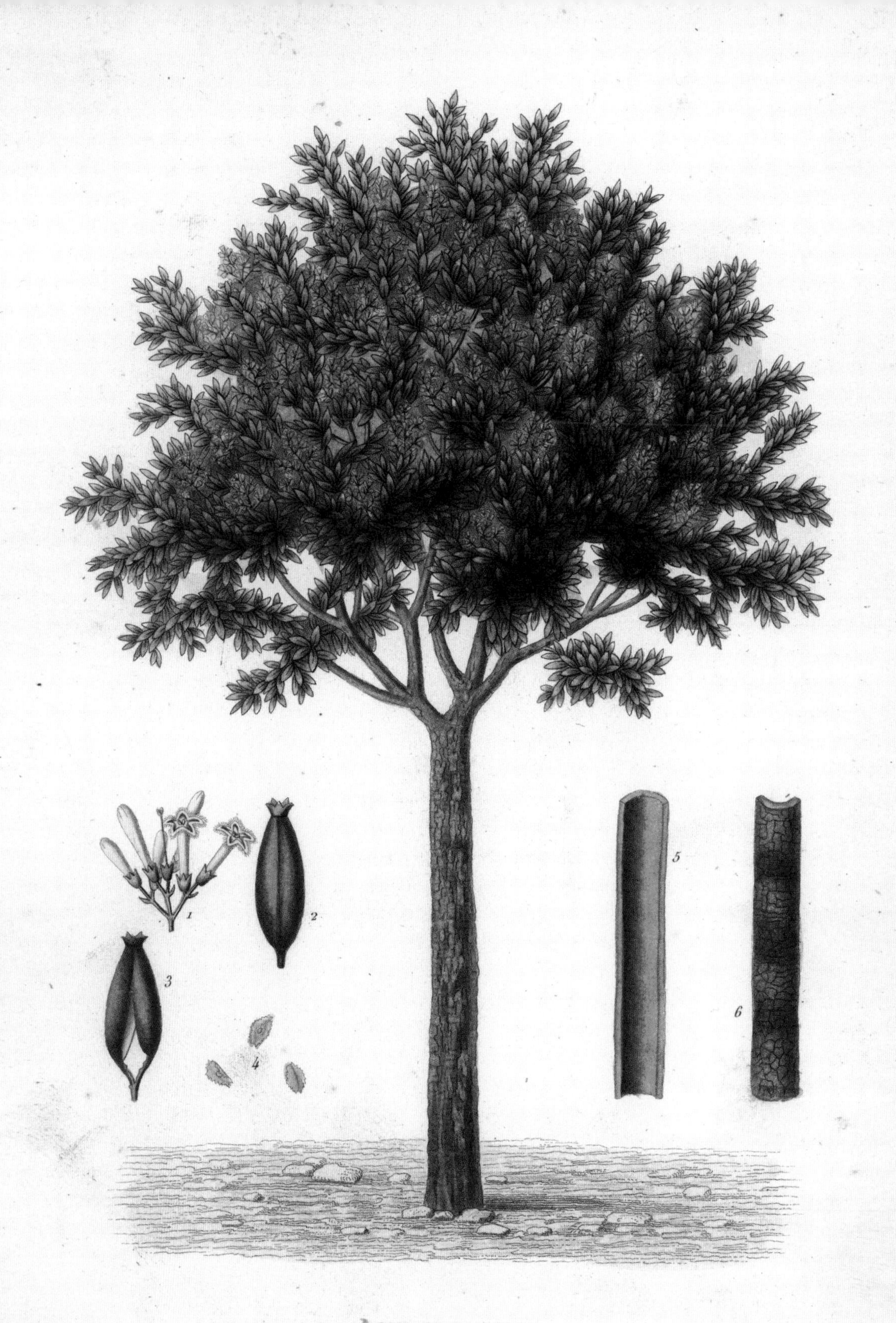

QUINQUINA GRIS,

Cinchona Condaminea, Humb.

PART I

THE TREE BEHIND THE TONIC

A NATURAL HISTORY OF TONIC WATER

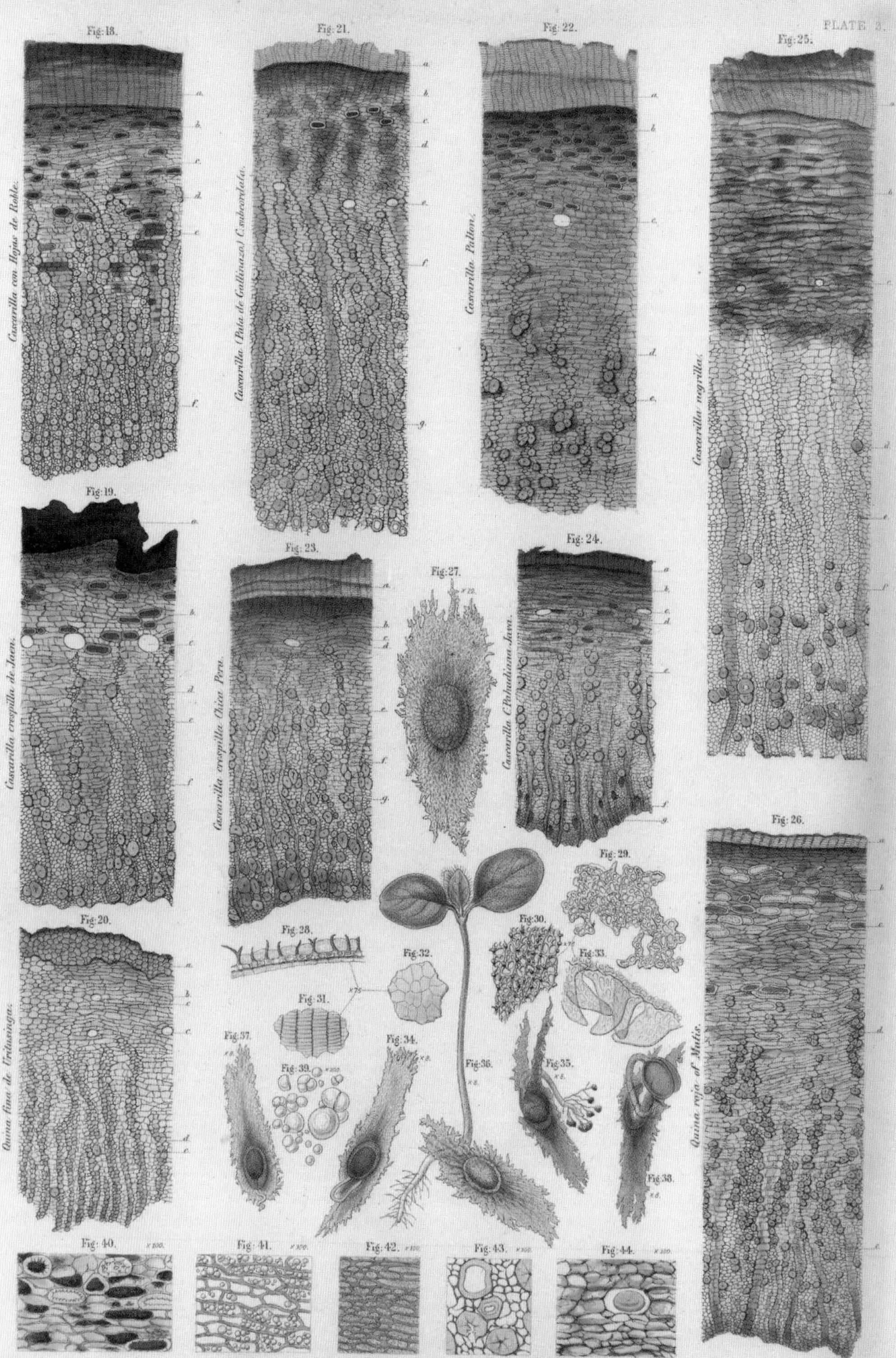

All magnified 50 diameters, except where otherwise marked.

I.

CINCHONA

THE MIRACULOUS FEVER TREE

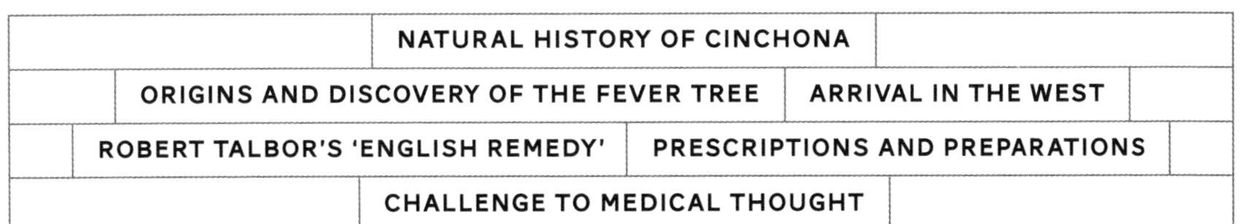

In the Peruvian Andes in 1847, French pharmacist Auguste Delondre saw a cinchona tree for the first time. He wrote of the experience: *'That magnificent tree. For so long I had seen it in my dreams and now it was before me. I lingered in ecstasy contemplating the silvery bark, those wide leaves in shimmering green and the flowers, such sweet perfume, somewhat reminiscent of lilac.'* Cinchona is indeed an attractive tree but the intensity of Delondre's response also reflects the fascination the tree has long held for pharmacists and botanists, the mystery that surrounded its origins, and the historic importance of its bark to medicine.

Cinchona bark contains quinine compounds that are powerfully antimalarial, one of which is the bitter flavouring in tonic water and bitter lemon drinks. However, for much of its history, cinchona bark remained mysterious. The active ingredients were not identified until 1820, malaria was not understood as a disease until eighty years after that and, crucially, most Western pharmacists and

[opposite] Microscopical observations of cinchona bark and seedlings by Walter Hood Fitch, from J. E. Howard, *Illustrations of the Nueva Quinologia of Pavon* (1862).

[previous page] Tree, bark, flower and fruits of grey cinchona (*Cinchona officinalis*). Barks were classified by traders and botanists as grey, white, yellow or red, according to their appearance and quinine content. Oscar Réveil, *Le Règne Vegetal: Flore médicale* (Paris, 1872). Florilegius / Alamy Stock Photo.

Heart of the Andes by Frederic Edwin Church in 1859, based on his travels in Ecuador two years earlier. Metropolitan Museum of Art.

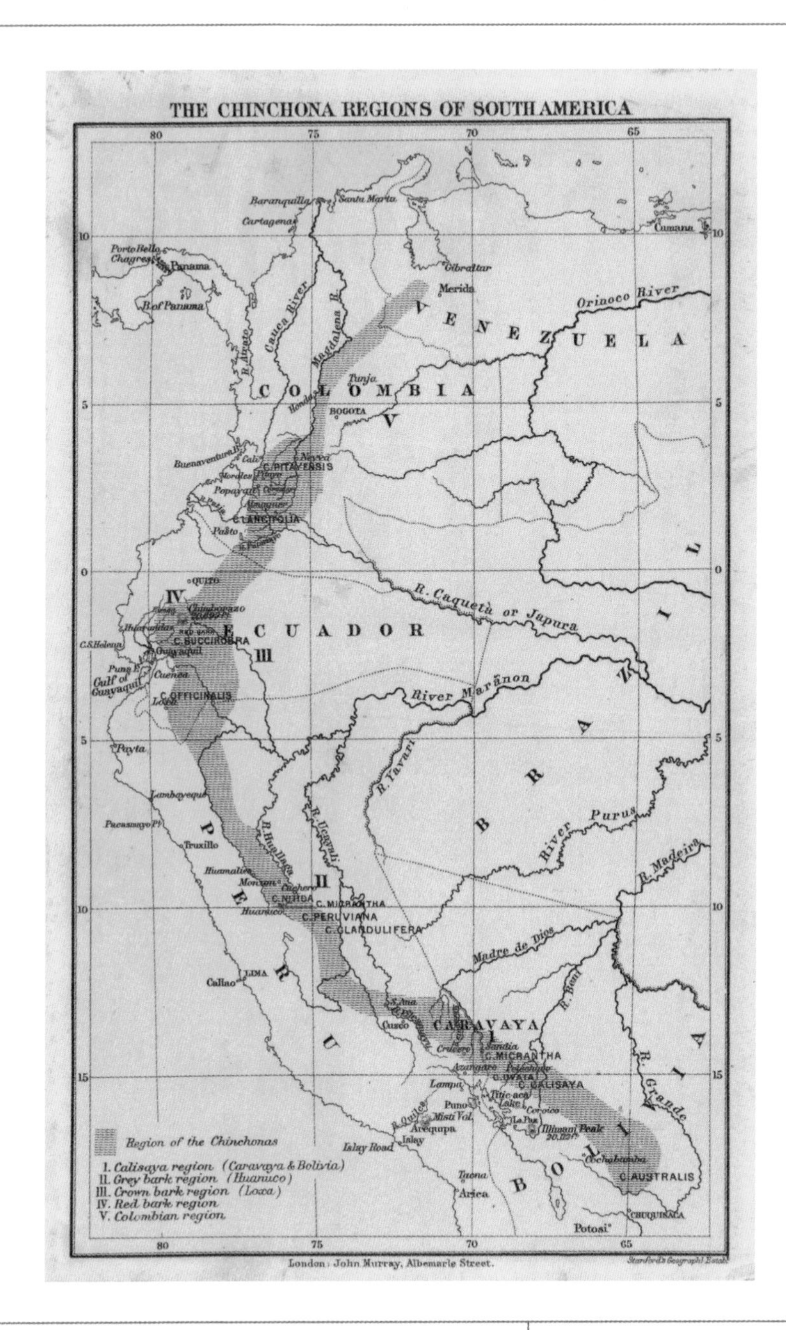

The shaded area indicates the native distribution of cinchona trees in the Andes. From *Peruvian Bark: A Popular Account of the Introduction of Chinchona Cultivation into British India, 1860–1880*, by Clements Markham (London, 1880).

botanists only saw the bark in the form it arrived by ship, as sacks of fragmented bark. These came from many different cinchona species growing in remote regions extraordinarily difficult for most to visit, on the eastern slopes of the Andes.

Malaria, the life-threatening parasitic disease, has been described as one of the biggest killers in history, with estimates that it may have killed half the people who have ever lived. Today it is thought of as a tropical disease, but until a century ago it was also the scourge of Europe. For over 350 years, the only effective treatment known to the West came from the bark of the cinchona tree, although Africa and China had their own antimalarial plants.

NATURAL HISTORY OF CINCHONA

Cinchona trees belong to the genus *Cinchona*, named by the Swedish botanist Linnaeus, which contains around twenty-five species, in the plant family Rubiaceae. Members of the family contain many medical and stimulant alkaloids such as caffeine from coffee (*Coffea* spp.) and dimethyltryptamine (DMT) from *Psychotria viridis* used in the psychotropic brew Ayahuasca, as well as the antimalarial quinoline alkaloids extracted from cinchona bark. These alkaloids have evolved to protect the plants from consumption by insects, browsing mammals and other pests. Some alkaloids act both through their taste – they are usually bitter – and through their poisonous effects. By chance, some alkaloids have desirable effects in the human body, but we should remember that many are in origin poisons. As we will see in Chapter 8, dosage is everything in the consumption of alkaloids, even when it comes to tonic water.

Cinchona trees grow along the eastern slopes of the Andes at an altitude of 1,000–2,500 metres (3,280–8,200 feet), in a narrow band running from north to south through Colombia, Ecuador, Bolivia, Peru and Chile. The damp and misty cloud forests in which cinchona resides are rich in plants, mosses, lichens and animals, a landscape that appeared exotic and captured the imagination of many visiting botanists and explorers. For example, the Monteverde cloud forest in Costa Rica

[above left] *Cinchona anglica*, a hybrid form that evolved in the Indian plantations. From *The Quinology of the East Plantations* by John Eliot Howard, illustrated by William Fitch (1869).

[above right] *Cinchona calisaya* (formerly *C. peruviana*), from *Illustrations of the Nueva Quinologia of Pavon* by John Eliot Howard, illustrated by William Fitch (1862). '*Its general aspect must be very pleasing, with its bright white and variegated bark, and elegantly formed and glossy leaves, changing towards their decadence into a rich scarlet.*'

has over 3,000 plant species, about twice the number of plant species native to the whole of the United Kingdom.

Cinchona trees are medium-sized, mostly up to twelve metres (forty feet) tall, with glossy evergreen leaves and loose clusters of fragrant white, pink, purple or red flowers. Their papery seeds are winged, contained in leathery capsules that split open and distribute them on the wind. It is hard to imagine that these diminutive seeds would cause such changes in the world when harvested by nineteenth-century explorers.

ORIGINS AND DISCOVERY OF THE FEVER TREE

Humans are endlessly ingenious in their use of the natural world. At least 30,000 species of medicinal plant have been used somewhere in the world, having been discovered by experimentation and through observing self-medication by wild animals. The origin of cinchona's medicinal use is, however, uncertain. Detailed records of medicinal plants used at the time of the Spanish conquest exist in the form of handwritten codices, but cinchona has proved stubbornly absent from these. The first definite reference to the use of cinchona bark for treating malaria was by Father Antonio de la Calancha, an Augustin priest based in Peru, who in 1633 recorded: *'[there] grows a tree which they call the fever tree (arbol de calenturas), whose bark, of the colour of cinnamon, made into powder and given as a beverage, cures the fevers and tertians; it has produced miraculous results in Lima'*. ('Tertians' are cyclical fevers that signify malarial attacks.)

There are reasons why cinchona's antimalarial properties might not have

Map of Villcabamba, Ecuador. The location of a legendary tribe, suggested in the 18th century to be the heartland of cinchona use. Wellcome Collection.

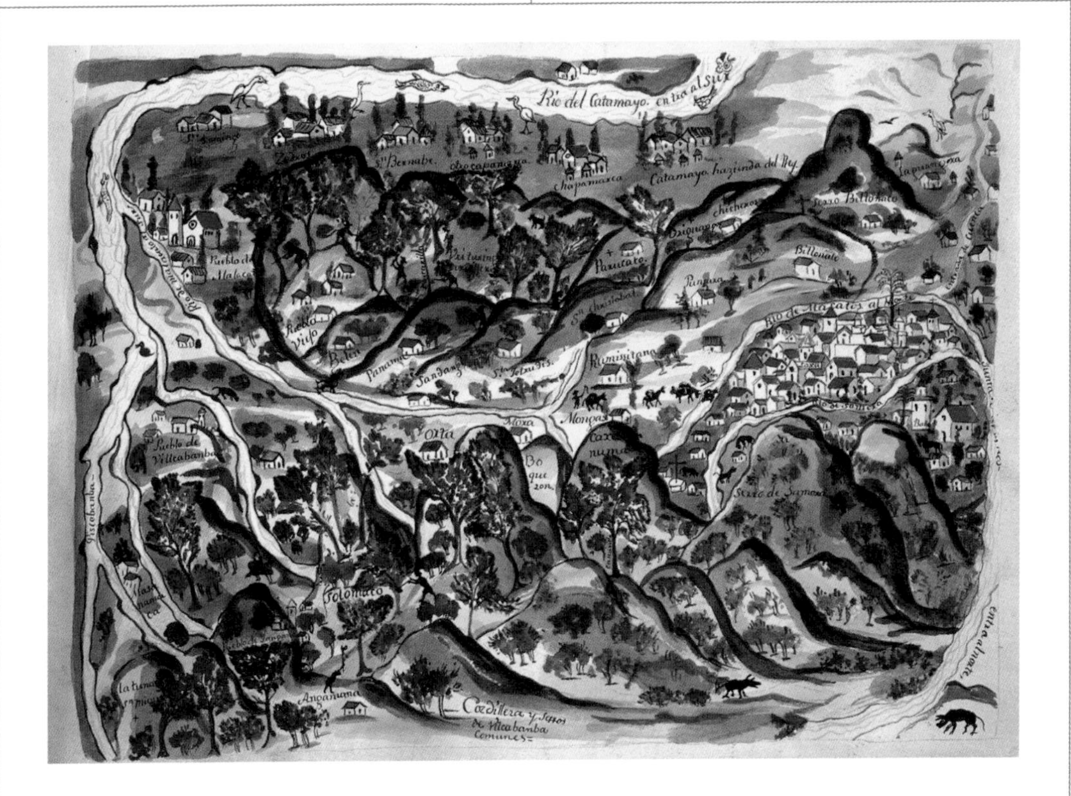

Fresco in the Ospedale di Santo Spirito in Rome, depicting the Countess of Chinchón being given cinchona. Wellcome Collection.

been discovered before the Spanish conquest. Severe forms of malaria were most likely absent from South America before the Spanish conquest; they arrived with enslaved people from West Africa and became rapidly established in low-lying coastal areas. In contrast, cinchona trees grow at high-altitudes, where malaria became prevalent as the railways brought increased numbers of people up the mountainsides in the nineteenth century. Thus, the trees with their quinine-rich bark and the malaria-ridden lowlands were geographically apart, making a chance discovery of cinchona's properties less likely.

Nonetheless, recent work by historians Fernando Ortiz Crespo and Matthew Crawford points to the high-altitude Loja region of Ecuador as a likely place of discovery of cinchona's properties, at least sixty years earlier than Calancha's record. Two intriguing references in books published in 1574 and 1600 refer to an unnamed tree that is strikingly similar in its properties to cinchona, exported from the port of Guayaquil. This serves the Loja region, renowned in the eighteenth century for the superb quality of its cinchona bark, and the source of bark for the Royal Pharmacy in Madrid. As well as having high-altitude cinchona and excellent trade connections, Loja was (and is) home to a very dynamic community of *curanderos*, or healers. It represents a highly probable place of discovery, but a definite answer awaits further discoveries in the Spanish archives.

Whenever it was discovered, the bitter bark of cinchona would have made it an excellent candidate as an antimalarial. Up until the end of the nineteenth century malaria was recognised by its pattern of fevers, as modern methods of diagnosis were unknown. Bitter plants have always been favoured for treating fevers. We now know that bitterness may indicate the presence of medicinal compounds including alkaloids such as quinine, or in the case of willow, salicin (an early forerunner of aspirin), which are often effective in treating fevers.

How the fever tree came to be used was as obscure in the seventeenth century

A cinchona forest in Latin America, from *Illustrated Travels: A Record of Discovery, Geography, and Adventure*, edited by H.W. Bates (1880).

The second illustration published in Europe of a cinchona tree. From *Dendrographias sive historiae naturalis de arboribus et fructibus tam nostri quam peregrini orbis libri decem*, by Johannes Jonstonus (1662). Wellcome Collection.

as it is today, so guesswork became legend. Stories invoked holy direction, jaguars and countesses. Some told of Jesuit priests who drank water from a lake into which some cinchona trees had fallen and were consequently cured of their fevers. French explorer and cinchona scholar, Charles Marie de La Condamine (1701–1774), wrote that local Ecuadorians had learned of the cure from watching mountain lions eat the bark. However, these stories pale in comparison to the legend of the Countess of Chinchon and her Miraculous Cure.

All good stories need a beautiful heroine, and Doña Francisca Henríques de Ribera, the eponymous countess and wife of the Viceroy of Peru, provides the back story for one of the most ubiquitous tales told about the discovery of cinchona, still told as 'fact' today. The story goes that as the beautiful countess lay languishing and feverish on her death bed in around 1630, she was given a dose of fever bark by either a devoted Peruvian maid or a concerned official of her husband. Her Miraculous Cure stirred her to return to Spain to philanthropically dispense the bark to the suffering people of Europe. Though charming, the story was eventually discredited by scholars who scoured her husband's journals which recorded, in meticulous detail, day-to-day life including family illnesses. At no point does he mention this otherwise remarkable event. In addition, the real Duchess died in Peru before ever returning to Spain.

It was the Countess of Chinchon that the founder of modern botany, Carl Linnaeus, was thinking of in his 1742 book *Genera Plantarum* when he gave the genus name *Cinchona* to the famous fever tree. However, all people, no matter how great, are open to error and Linnaeus is no exception. He misspelled the countess's name and the cinchona was ever after known without the first 'h', causing confusion over pronunciation to generations of botanists ever since.

ARRIVAL IN THE WEST

The Cinchona bark cure was sold under a range of names including fever tree, quina, calisaya (a term of the indigenous Aymara peoples of the Andean area meaning 'best bark'), Peruvian bark and Jesuit's bark. It was sometimes graded by colour – the best bark was said to be the red bark, lesser barks being yellow and grey. By the 1640s, parcels of harvested bark were imported to Europe, but between the distant trade ports of the Americas and Europe much was lost in translation. The bark was often confused with that of

A pharmacist grinding herbs in France, by J.I. Grandville (c. 1840). Wellcome Collection.

another medicinal tree, the quinaquina or Peruvian balsam (*Myroxylon balsamum*). With uncertainty over identification as well as dosage, it is unsurprising that 'cinchona' barks were unreliable cures and took time to be proven to be as miraculous as claimed.

There was also a cultural barrier to the acceptance of cinchona as a medicine in Britain. On the stage of an oft-warring Europe, rivalry between countries driven by religious tensions meant that medicines promoted by one section of society could be treated by suspicion by another. With a name like Jesuit's bark reflecting the Catholic advocates and exporters of the medicine, countries of other denominations, particularly the

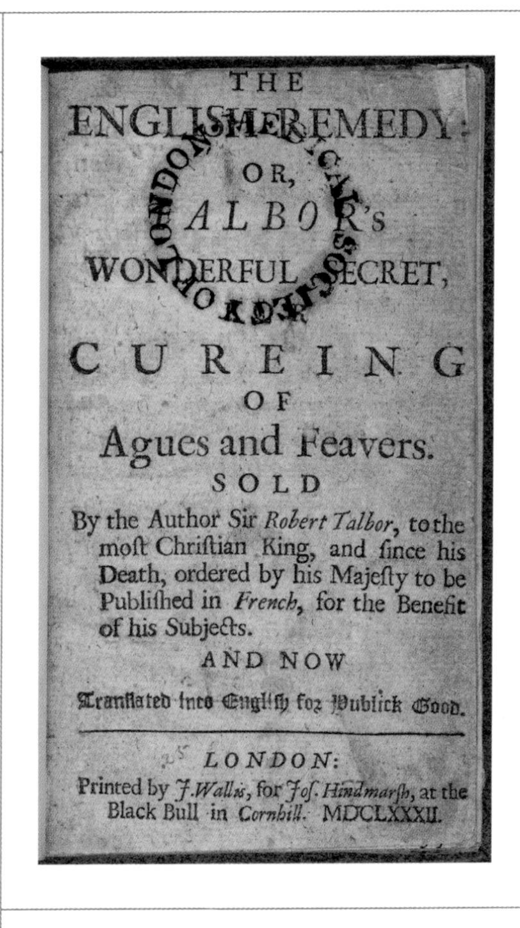

THE
ENGLISH REMEDY:
OR,
TALBOR'S
WONDERFUL SECRET,
CUREING
OF
Agues and Feavers.
SOLD
By the Author Sir *Robert Talbor*, to the moſt Chriſtian King, and ſince his Death, ordered by his Majeſty to be Publiſhed in *French*, for the Benefit of his Subjects.
AND NOW
Tranſlated into Engliſh for Publick Good.

LONDON:
Printed by *J. Wallis*, for *Jos. Hindmarsh*, at the Black Bull in *Cornhill*. MDCLXXXII.

Talbor's 'English Remedy', commissioned by King Louis XIV, 1682. Wellcome Collection.

Protestant English, resisted its use. This was compounded by Jean-Jacques Chifflet (1588–1660) who wrote a critique of cinchona bark as a 'Popish fraud'. Even Oliver Cromwell is said to have refused the 'Popish powder' – before he subsequently died of malaria.

However, a remedy so effective could not stay unpopular for long and it eventually came into common use all over Europe. By 1677 Peruvian bark was recorded as *'an excellent thing against all sorts of ague'* (agues are the chills that accompany fevers) in the official *Pharmacopoeia Londinensis* of the Royal College of Physicians, but its popularity in Europe from this time is credited to the efforts of a few key figures: English physician Thomas Sydenham (1624–1689), known as the English Hippocrates, who was an early advocate of the bark (see p. 33), Italian physician Francisco Torti (1658–1741) who specified for which type of fevers it should be used (see p. 32), and Robert Talbor who brought it in to popular use.

ROBERT TALBOR'S 'ENGLISH REMEDY'

Cambridgeshire-born Talbor (1642–1681) originally trained as an apprentice apothecary before heading to university. There, he developed an interest in the study of fevers and moved to the Essex marshes in 1668 where *'the agues are the epidemical diseases'*.

He experimented with cures, and eventually developed a secret recipe for fevers. In an astute move, he publicly opposed the *'misuse'* of cinchona bark and instead promoted his own remedy as a *'noble and safe'* alternative. His recipe remained a secret and was simply described as containing *'two foreign and [two] domestick'* vegetables. It was only a fortunate chance that cast his recipe into the limelight. While treating feverish soldiers visiting in the marshes, Talbor cured a French officer who had the ear of King Charles II. Talbor was invited to court and after curing the king, he was knighted in 1678 and his fame was secure. Charles sent him to France to treat a relative, and there he gained the notice of King Louis XIV. Louis wanted access to the secret of the marvellous recipe and Talbor agreed, on condition that it would not be revealed until his death. Sadly, this came not long after; Talbor returned to England in 1681 to Cambridge University but died in the same year. His recipe was published shortly after, on the order of King Louis, under the title *The English Remedy or, Talbor's Wonderful Secret, for Cureing of Agues and Feavers. Sold by the Author Sir Robert Talbor, to the most Christian King, and since his Death, ordered by his Majesty to be Published in French, for the benefit of his Subjects. And now translated into English for the Publick Good.*

It was then revealed that Talbor's successful remedy had included Jesuit's bark all along. It may have been the manner of preparing it which made it particularly effective; he used an infusion as well as an alcoholic tincture, which would have allowed a range of chemical constituents to have been extracted. Talbor had taken half a kilo (about one pound) of powdered bark, infused it with parsley and aniseed, and then added it to approximately seven litres (fifteen UK pints) of wine. He occasionally served it with additional lemon, orange and rose flavourings, which may have masked the taste of the bark from reluctant patients. The recipe also included pain-relieving opium. Cinchona's place in medicine was firmly established and the 'English Remedy' became a popular preparation throughout Europe. During the eighteenth century, imports of cinchona dominated the drugs trade, making up forty per cent of the total drugs imported to England from the Americas.

PRESCRIPTIONS AND PREPARATIONS

Mixing other herbs with the bitter, powdered Jesuit's bark was not unique to Talbor – it was common practice in early preparations. The bark arrived in Europe in dried, stripped pieces, to be ground to a powder ready for use. It could be made into pills or dissolved in port wine, with its bitter taste masked by flavoursome herbs such as cinnamon, cloves and orange peel, and sweet syrups made with treacle or honey. Recipes commonly included purgatives and emetics, which were believed to remove illness in the body. Purgatives included senna (*Senna alexandrina*) and jalap (*Ipomoea purga*), as well as other bitters such as oak bark. As in Talbor's recipe, pain-relieving opium was sometimes added to soothe restless patients. There are references in Britain to the use of *Artemisia* species such as mugwort *(A. vulgaris)* and southernwood *(A. abrotanum)* for treating agues. This is interesting as it shows a use similar to the closely related Chinese wormwood (*A. annua*), which was recorded for anti-fever properties in an 2,000-year-old Chinese herbal. Extracts of Chinese wormwood are now used as a modern antimalarial drug (see p. 38), though it is unclear if the levels in the British varieties would have been high enough to have much effect.

CHALLENGE TO MEDICAL THOUGHT

The success of cinchona significantly changed beliefs about medicine in Europe. The Galenic medicinal system (based on the works of the second-century Greek physician Galen) taught the Hippocratic view that the body is governed by four humours, with imbalance of these causing illness. The use of purgatives, emetics and bleeding would help to move the humours and return the patient to good health. Herbs were used symptomatically and their effect on the body was harnessed to produce the opposite effect from the illness at hand. Therefore, feverish patients should be treated by the application of cooling remedies to counteract the body's heat. The bitterness of cinchona bark meant that it was classed as a heating remedy – a significant challenge to Galenic doctrine that delayed its wider adoption. However, its swift curative effects were undeniable and encouraged new experimentation and focus on the active ingredients of medicines.

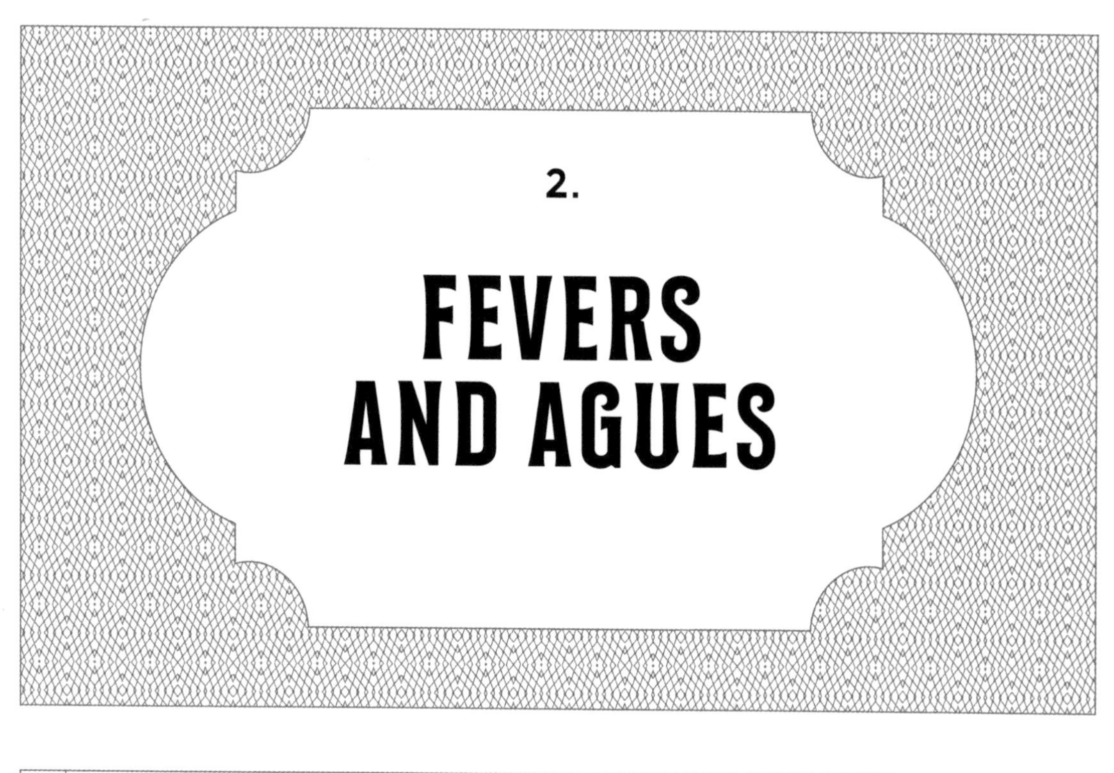

2. FEVERS AND AGUES

Malaria has been around since before our earliest human history. The oldest known malarial parasites have been found, Jurassic-Park style, in the body of an amber-preserved mosquito dating from thirty to forty million years ago. Scientists believe that malaria spread to humans from apes in Africa, co-evolving with our human ancestors over the last 10,000 years.

Most Europeans think of malaria as a tropical disease, but one hundred years ago it would also have been a concern to many people living in cooler areas, including marshy parts of Britain such as the counties of Essex and Kent. At one point, the disease reached almost to the Arctic Circle. Though the Western European strain of malaria is not quite as severe as the one found in tropical areas, it made many people sick, weakening them over time and making them susceptible to other diseases.

MALARIA AND MOSQUITOES

'If you ever thought that one man was too small to make a difference, try being shut up in a room with a mosquito.'
(ascribed to the Dalai Lama)

[opposite] Suffering from malaria, from *Kranken-Physiognomik* by K. H. Baumgärtner (1929). Wellcome Collection.

Malaria is caused by a tiny single-celled parasite of the *Plasmodium* genus that inhabits the blood cells and organs of its host. There are four main *Plasmodium* parasites that cause malaria in humans, each slightly different in symptoms and severity. The most serious and frequently fatal illness is caused by the species *Plasmodium falciparum*. This is found in the subtropics and tropics worldwide and predominates in Africa. The *P. vivax* species is most common and leads to a milder form of malaria; it is dominant in Asia and Latin America, and was once the species found in Europe along with *P. malariae*.

For *Plasmodium* to spread between infected persons to new hosts, a vector is needed – a role undertaken by various species of the diminutive yet troublesome bloodsucking mosquito in the genus *Anopheles*. While mosquitoes mostly rely on a diet of plant sap and nectar, once the females are ready to breed they seek out protein-rich blood to nourish their eggs – and humans make an ideal meal. When sated, the female mosquito needs standing water to lay her eggs in, and puddles, ditches, ponds and marshes provide the perfect breeding grounds. The larvae hatch within a few days, munch on algae, then emerge as fully developed mosquitoes ready to repeat the cycle.

MALARIAL CONTROL

Tackling malaria is as much about controlling the habitat of mosquitoes to interrupt their life cycle, as it is about treating the disease itself. Draining marshes and covering and treating water are all fundamental steps for effective control. Other measures include preventing mosquitoes from biting, particularly during their active night period, through bed nets and

[above left] Mosquito headnet with cap and 'breathing' pipe, early 20th century. Wellcome Collection.

[above right] Annaba, Algeria: power spraying the exterior of thatched huts to control mosquitoes (1944). Wellcome Collection.

[opposite] Calendar with antimalarial advice, given to American GIs in the South Pacific. Artist Frank Mack (1945). National Library of Medicine.

insecticides. DDT was found to be a highly effective agent against mosquitoes in the 1950s and 1960s. However, its use has reduced, owing to its damaging effects on other insects, birds and humans, and to the evolution of resistance to DDT in mosquitoes.

The drainage of fens and marshes, as well as the development of closed sewer systems, helped eradicate malaria in Britain and other areas of Europe. A temporary resurgence occurred during the Second World War when malaria-bearing soldiers stationed in tropical areas returned to Britain. Today, the relatively few cases of malaria recorded in Europe can be traced to patients travelling from the tropics.

In the battle between humans and *Plasmodium*, some forms of natural immunity have developed. Sickle cell disease and thalassaemia affect many with ancestry from Africa and southern Europe. Sickle cell disease is named after the sickle-shaped red blood cells it causes;

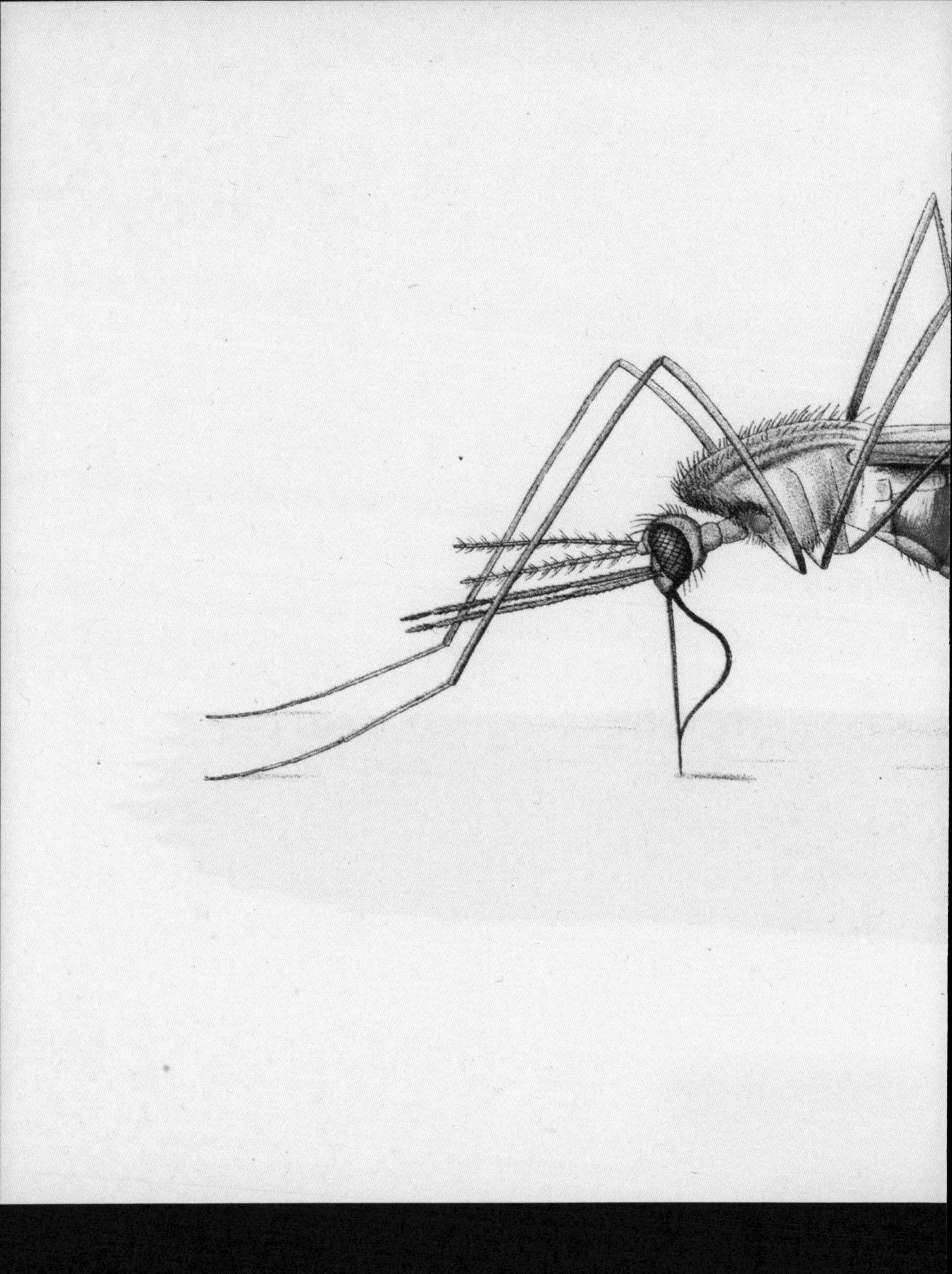

Anopheles maculipennis, the mosquito responsible for transmitting malaria in Europe, from *The Journal of Hygiene* (1901). Wellcome Collection.

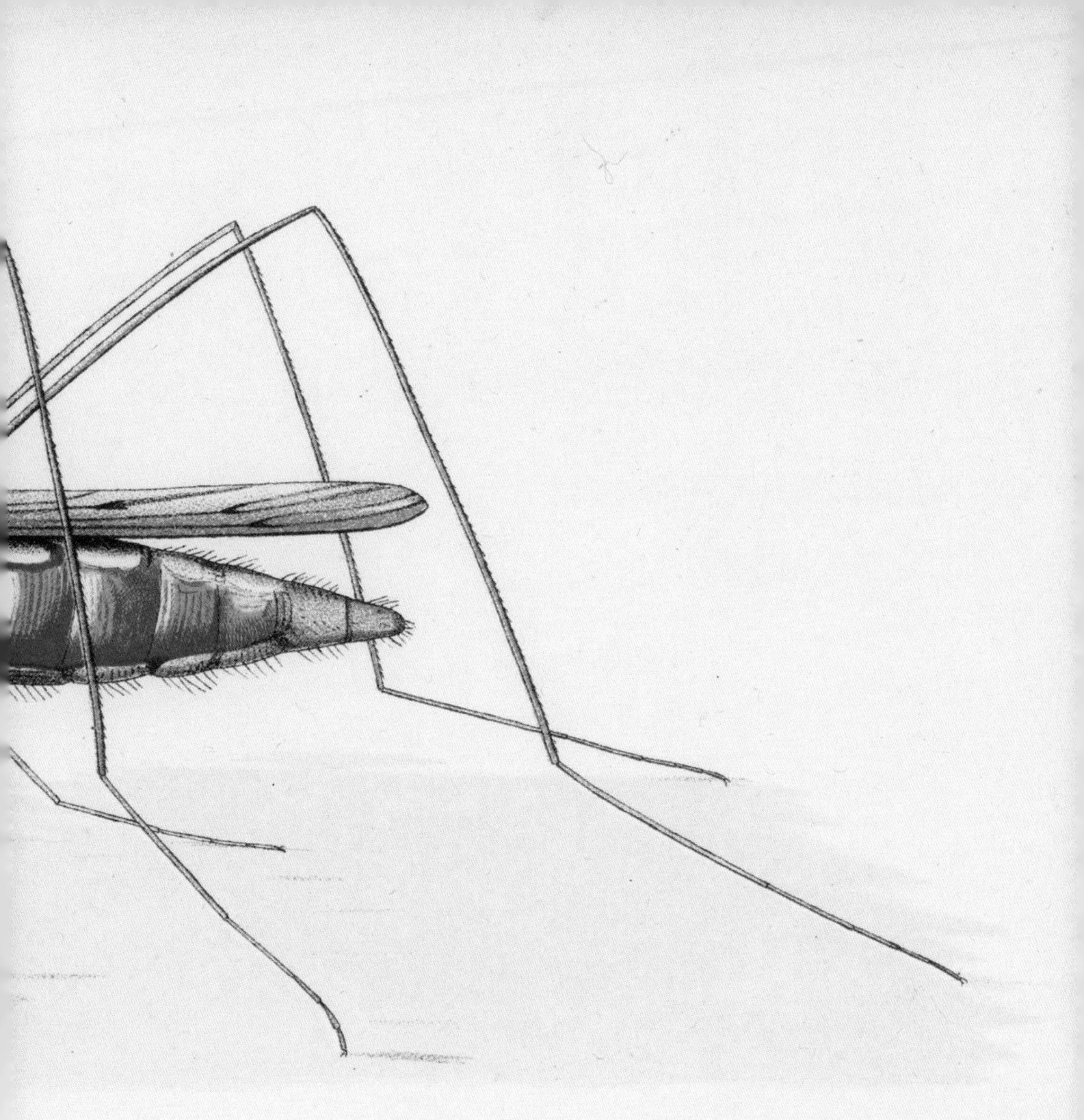

ANOPHELES MACULIPENNIS. ♀

SUCKING BLOOD

E. Wilson, Cambridge.

SYMPTOMS OF MALARIA

Uncomplicated: intermittent fever, chills, profuse sweating, aches and pains, anaemia, headache, abdominal pain, nausea and vomiting
Severe: enlarged spleen, jaundice, severe anaemia

thalassaemia is a group of conditions that affect the amount of red blood cells produced, resulting in anaemia. These inherited disorders cause some serious health problems but they also convey some immunity to malaria, and have thus evolved as an advantage to populations in malarial areas.

MALARIA IN HISTORY

Although the mosquito-parasite cause of malaria was not understood until the end of the nineteenth century, fevers were long a well-known symptom of illness. During the late eighteenth century, French physician Félix Vicq d'Azyr classified 128 separate fever types. Among these, a distinct set of cyclical feverish symptoms allows malaria to be identified in historical sources. These typically refer to fevers and 'agues', the chills and shivering that occur before the fever, or 'intermittents', in which fevers are classified by their repetitive cycle.

These patterns were given descriptions: 'quotidian' for a cycle repeated every twenty-four hours, 'tertian' for one repeated every first and third day after incubation; and 'quartan' for one repeated every first and fourth day. It is now known that these cycles are caused by the varied timings in the life cycle of different malarial parasite species. After the parasites reproduce, they burst out of their blood cell hiding places to start a new cycle of infection. The body senses the presence of the parasites released into the blood stream and responds with fever. *Plasmodium malariae* has a quartan cycle with a fever every third day; the other *Plasmodium* species have tertian cycle with a fever every second day. If contracted in the autumn, malaria can also 'hibernate' over winter, re-emerging in the spring, causing what was called spring fever, another term that included malaria.

[above] A) Soldier, take your quinine every day: He didn't take his quinine/ He took his quinine. B) Always sleep under a mosquito net. On waking: This man had no mosquito net/ This man slept soundly under his mosquito net. Postcards by A. Guillaume (1914–18). Wellcome Collection.

[opposite] Tree of fevers, from *Therapeutice Specialis* by Francisco Torti (1712). Wellcome Collection.

LIGNUM FEBRIUM.

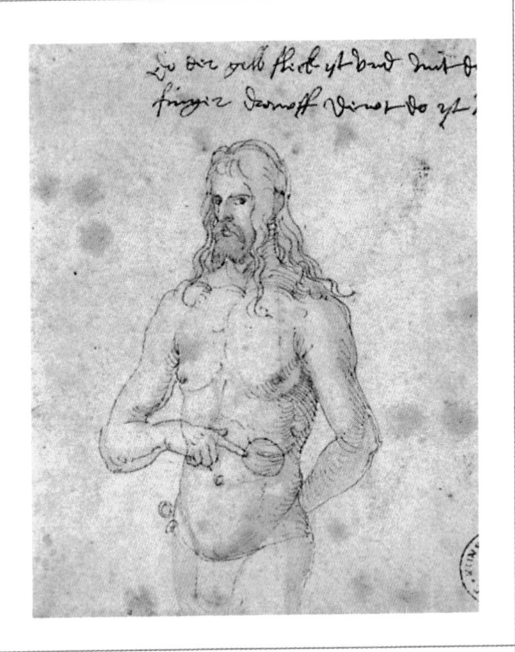

Chinese physicians recorded an illness believed to be malaria in a medical book dating to the first century AD, the *Huang Di Nei Jing* [*Yellow Emperor's Inner Canon*], discussing the relationship between fevers and enlarged spleens. The spleen, which removes infected blood cells, becomes overwhelmed in malaria, resulting in painful swelling. The sixteenth-century German artist Albrecht Dürer drew a self-portrait, pointing to the area of the spleen during a bout of what was probably malaria, captioned '*Do wo der gelb fleck is und mit dem finger drawff dewt do is mir we.*' [*There, where the yellow spot is located and where I point my finger, there it hurts.*

In 1712, the Italian physician and fever specialist Francesco Torti (1658–1741) published a beautiful family tree of fevers, the *Lignum februm*, a stylised cinchona tree showing, along its branches, different forms of fever and their relationships. The tree also shows, through the contrast of a healthy leafy side with a withered, parched one, the responsiveness of certain fevers to cinchona bark, and that it is the *intermittent fever* (repeated, cyclical) type that is treated by cinchona bark. This tree of fevers helped to clarify which would respond to cinchona – in other words, which were malarial – leading to more consistent treatment. By the mid-eighteenth century, cinchona was widely accepted as a reliable treatment for intermittent fevers.

HISTORIC TREATMENTS

Before the discovery of cinchona, herbs used in the treatment of fevers included those used to reduce and to cool heat, including gentian (*Gentiana lutea*), agrimony (*Agrimonia eupatoria*) and barley water (*Hordeum vulgare*). In ancient Rome patients were given amulets, some with *abracadabra* written on them. Holy relics and charms persisted well into the medieval period to ward off sickness. However, prayers, chants and herbs were of little use against severe malaria, so it is little wonder that cinchona became a wonder-drug.

MARSHES AND MIASMAS: THE CAUSES OF MALARIA

Early medical beliefs, originating from the ancient Greek Hippocratic corpus and surviving into the nineteenth century, attributed the cause of some illnesses to the theory of miasma. This held that diseases were formed from noxious 'exhalations' in the air, often associated with areas of bad smells such as marshy districts. In some ways, this was not far from the truth – mosquitoes favour marshy areas for breeding, so people living near marshes were more affected. The area around the Roman Campagna was infamously malarial and marshy, and many travellers to Rome were struck down with a virulent form of 'Roman fever'. The word malaria derives from the Italian for 'bad air', and was introduced into English by Horace Walpole in 1740 when he wrote '*there is a horrid thing called the malaria, that comes to Rome every summer, and kills one*'.

Few links between fevers and insects can be found in early records, though notably the *Susruta Samhita*, an ancient Ayurvedic text written sometime between 250 BC and AD 500 by the physician Susruta, discusses the link between insects and various types of fever. Over a thousand years later Thomas Sydenham (1624–1689), an influential English physician, was an early advocate for the use of cinchona bark. In his *Observationes Medicae* of 1667 he described the link between those living near marshes and fevers, noting that '*when insects do swarm extraordinarily and when…agues (especially quartans) appear as early as about midsummer, then autumn proves very sickly*'. This link would remain unexamined for another 200 years, when the parasite and the mosquito were finally revealed as the troublesome duo.

[above] The symptoms of malaria imaginatively rendered by James Dunthorne and Thomas Rowlandson (1788). Wellcome Collection.

[opposite] *The Sick Dürer* by Albrecht Dürer (1471–1528), at the Kunsthalle, Bremen. Wikimedia Commons.

HOW QUININE WORKS

Quinine's action is not fully understood but it is thought that it causes a complex molecule called haemoglobin to become toxic to the *Plasmodium* parasite while it inhabits red blood cells. *Plasmodium* digests the globin parts and packages up the toxic heme part into a harmless pigment called hemazoin. Quinoline alkaloids such as quinine interrupt this stage, and the parasite cannot process the toxic heme out of its body, essentially poisoning itself.

LABORATORIES AND MICROSCOPES: NINETEENTH-CENTURY BREAKTHROUGHS

The first breakthrough in understanding malaria came in 1880 when Charles Louis Alphonse Laveran (1845–1922), a French Army physician based in Algiers, observed a small wriggling parasite – *Plasmodium* – in the blood of a feverish soldier. Though particles had been observed in the blood of malarial patients in 1858, no one had observed that they could move and were, in fact, parasites. Laveran won the Nobel Prize for Medicine in 1907 for his work. A cascade of further discoveries followed. In the mid-1880s, the Italian Camillo Golgi went on to discover that the parasites divided at regular intervals that coincided with the intermittent fever cycles, and Ettore Marchiafarva and Angelo Celli worked out the life cycle of the parasite.

The mosquito as the final link in the spread of malaria was uncovered simultaneously by India-based physician Ronald Ross (1857–1932), and Italian Giovanni Battista Grassi (1854–1925),

[left] Susruta explains his theory to a potentate. Wellcome Collection.

[opposite left] Sir Ronald Ross, Mrs Rosa Ross and Mahomed Bux, Ross's laboratory assistant, with bird cages on the steps of laboratory in Calcutta (1898). Wellcome Collection.

[opposite right)] Giovanni Grassi (1854–1925). Wellcome Collection.

A STUDY IN A SICKROOM

While a young journalist in India in the 1880s, Rudyard Kipling often suffered from malarial fevers. This compelling account, written in Simla in 1885, vividly describes the effects of malaria and a heavy dose of quinine:

A brisk canter in May on a pulling horse; violent perspiration, followed by a twenty minutes' lounge at the public gardens, where the flooded tennis courts reek like so many witches' cauldrons, and the Enemy is upon you. Neither Mrs Lollipop's banalities, the maturer charms of the Colonel's wife, nor the fascinations of a gin and tonic at the peg table will keep him at bay. With the dreary foreknowledge, born of many previous experiences, you shall recognize that, for the next twelve hours at least, you are 'in for it'; and shall communicate the fact with a sickly smile to your friends. The instinct of the stricken wild beast for rest and retirement drives you to your bachelor quarters. Man's wisdom recommends quinine and an early retreat bedward.

You were to have dined out to-night, and by this time should have been in your trap on the way to Mrs Lollipop's. But man proposes and the fever disposes. You have sailed far out of the reach of such mundane matters as dinners and flirtations, and are alone in that strange phantasmal world that lies open to us all in time of sickness — on the first stage of your journey towards the Purgatory of sizes and distances. Of this you are dimly conscious, for the racking pains in legs and trunk have given place to pains in the eyes and head only. The cold fits have passed away, and you have been burning steadily for the last ten minutes, preparatory to a final glissade down a rolling bank of black cloud and thick darkness, and out into the regions beyond. Here you are alone, utterly alone on the verge of a waste of moonlit sand, stretching away to the horizon. Hundreds and thousands of miles away lies a small silver pool, not bigger than a splash of rainwater. A stone is dropped into its bosom, and, as the circles spread, the puddle widens into a devouring, placid sea, advancing in mathematically straight ridges across the sand. The silver lines broaden from east to west, and rush up with inconceivable rapidity to the level of your eyes. You shudder and attempt to fly. The innumerable lines retreat with a long drawn 'hesh-sh' across the levels, and the terrible sea is contracted to the dimensions of a little puddle once more. A moment's breathing space, and the hideous advance and retreat recommences. The unstricken observer would tell you, if you cared to listen (which you do not, for you are deep in a struggle for life), that this phenomenon is simply the result of the quinine taken a few hours ago.

A NEW FEVER DRUG

While searching for an alternative to cinchona growing closer to home, Oxfordshire priest Edward Stone (1702–1768) discovered another fever drug. In 1763, he reported that he had '*accidentally tasted*' willow bark (*Salix* sp.) six years previously and '*was surprised at its extraordinary bitterness, which immediately raised me a suspicion of its having the properties of the Peruvian bark*'. Using the old medical reasoning of the 'doctrine of signatures' – where a cure could be signified by a clue in the form or habit of a plant – he noted that as willow grows in damp places, typical of areas where fevers are most prevalent, the tree must provide the cure. This was a rediscovery of older knowledge, as Hippocrates had promoted its use for fevers too. Stone was correct to some extent, although willow wasn't a treatment for malaria, it was beneficial for treating some fevers. The active compound, salicin, was later modified and marketed as aspirin, a powerful medicine for pain, inflammation and fever.

who published their discoveries just four months apart. Ross worked on bird-borne malaria and the *Culex* mosquito. In 1897 he observed the malaria parasite burst in the dissected stomach of the mosquito. The resulting 'threads' migrated to the salivary glands, ready for delivery to the next unfortunate host. Grassi confirmed that it was the *Anopheles* genus of mosquito that was the vector in humans. Ross subsequently won the Nobel Prize for Medicine in 1902, controversially as many believe it should have been shared with Grassi.

[above left] Charles Alphonse Laveran (1845–1922) spearing a mosquito. Wellcome Collection.

[above right] Life cycle of the malaria parasite from *Die Malaria, Studien eines Zoologen* by Giovanni Battista Grassi (1901). Wellcome Collection.

[opposite] Wartime poster showing the link between mosquitoes and malaria (1941). Wellcome Collection.

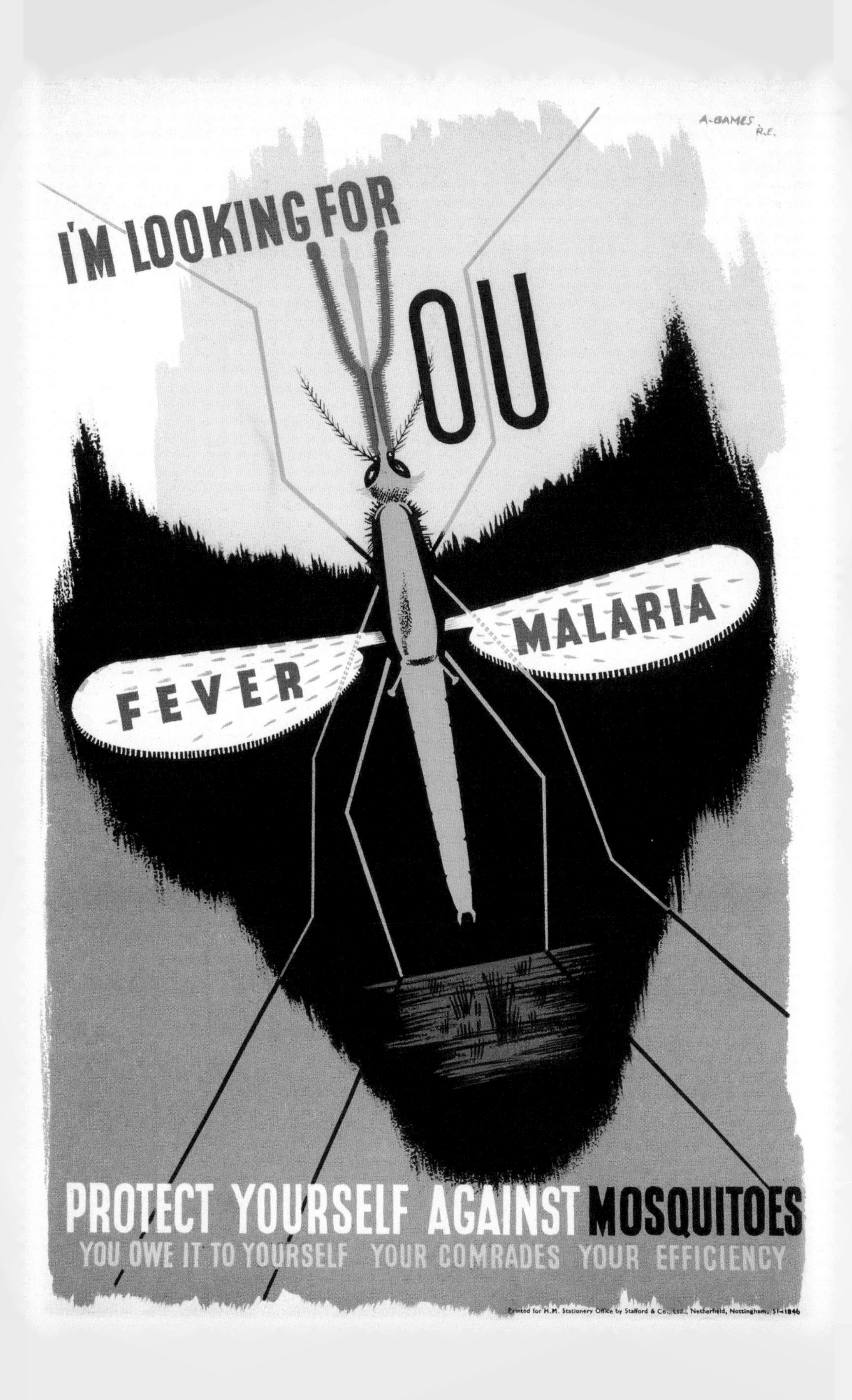
A-GAMES R.E.
I'M LOOKING FOR YOU
FEVER
MALARIA
PROTECT YOURSELF AGAINST MOSQUITOES
YOU OWE IT TO YOURSELF YOUR COMRADES YOUR EFFICIENCY
Printed for H.M. Stationery Office by Stafford & Co., Ltd., Netherfield, Nottingham. 51-1846

MALARIA TODAY

Despite medical advances, malaria remains one of the biggest health problems in over ninety countries. According to the World Health Organization's 2018 malaria report, there were 219 million cases of malaria that year and 445,000 fatalities, mostly of young children. The total global spend by governments tackling the issue during 2017 was a huge US$2.7 billion, with a huge impact on economic development, particularly in sub-Saharan Africa. Though incidences of malaria have been dropping each year, owing to increased access to medicines and control of vectors, researchers are predicting that rising temperatures from global warming may lead to a return of malaria into areas from which it has been eradicated, including Europe.

[opposite] Sweet wormwood (Qing Hao, *Artemisia annua*) from *Diannan Bencao Tushuo* (*The Illustrated Yunnan Pharmacopoeia*), manuscript of 1773. Wellcome Collection.

[below] Tu Youyou, Winner of the 2015 Nobel prize for Medicine for her part in the discovery of the antimalarial compound artemisinin extracted from *Artemisia annua*. Wikimedia Commons.

QING HAO: A CHINESE HERB

During the 1960s chloroquine and other related quinolines, the synthetic replacements for quinine developed in the 1940s, met a growing problem of resistance in the malarial parasite, like that which had emerged earlier for quinine. During the Vietnam War, both the major countries involved, China and the United States, looked for alternative antimalarials. The Chinese team operated under difficult conditions as the Cultural Revolution (1966–76) targeted researchers. The project tested many herbs without success but homed in on the drug Qing Hao, listed as a fever treatment in many works on Chinese medicine, the earliest being *A Handbook of Prescriptions for Emergencies (Zhou Hou Bei Ji Fang)* by Ge Hong, written about AD 400.

Results of animal trials with Qing Hao – several species of wormwood (*Artemisia*) – were disappointing. The team leader, Tu Youyou, re-read the prescription in the *Handbook of Prescriptions* and noticed that, unusually for a herbal decoction, it did not recommend heating the water in which the herb was to be extracted. In 1971 she tried laboratory extraction at a lower temperature and the herb immediately gave improved results. In 1972 the active ingredient, artemisinin, was extracted and identified, and found to be in greatest quantity in *Artemisia annua*. After many years of clinical trials, use of drugs derived from artemisinin began in the 1990s, with the first resistance in malarial parasites appearing in 2008. Artemisinin has had a hugely positive effect on malaria treatment, and the World Health Organization runs a programme to ensure that different variants of the drug are used, in combination with other types of antimalarial drug, to slow the development of resistance. In 2015 Tu Youyou was awarded the Nobel Prize for Medicine for her work.

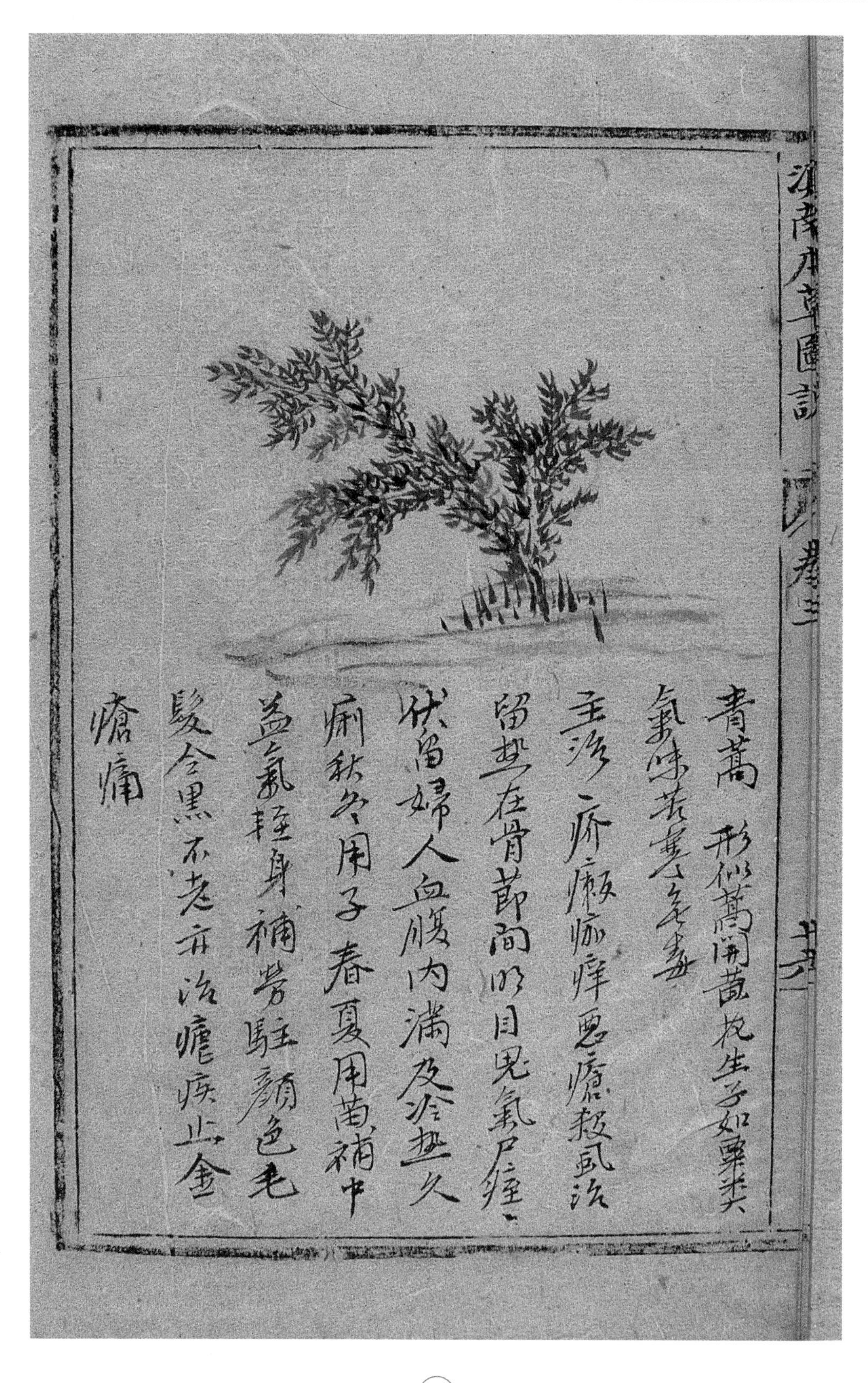

滇南本草圖說　卷三　十六

青蒿　形似蒿開黃花生子如粟米大
氣味苦寒無毒
主治疥瘙痂癢惡瘡殺虱治
留熱在骨節間明目鬼氣尸疰
伏留婦人血腹內滿及冷熱久
痢秋冬用子春夏用苗補中
益氣輕身補勞駐顏色毛
髮令黑不老亦治瘧疾止金
瘡痛

3.

PLANTATIONS AND POLITICS

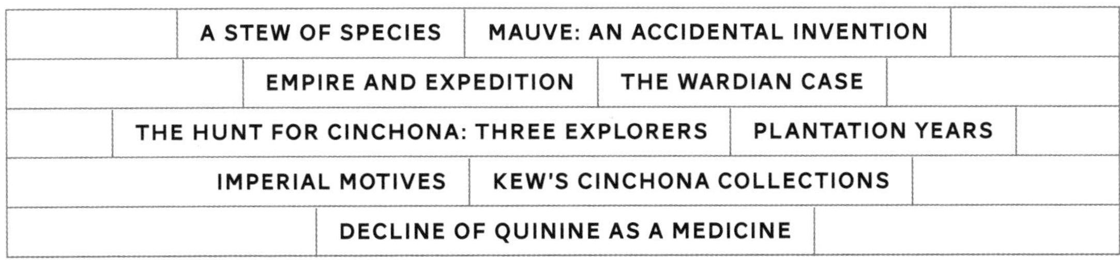

Harvesting cinchona involved trekking deep into the Andean forests, finding scattered clumps of trees, chopping them down and stripping them of bark. Nineteenth century botanists presented this as unsustainable harvesting, but it is likely that coppicing was practiced, in which cut trees re-grow vigorously from the roots. Similar techniques were later to be applied in the Indian plantations. There were, in addition, state-sponsored efforts to plant new trees for those felled. However, during the early nineteenth century the political situation in South America became unstable, as nations asserted their independence from Spain. In 1844, Bolivia brought in measures to control cinchona harvesting but failed to pay workers proper wages and a black market in barks arose. European empires urgently desired better control over quantity, quality and price, and expressed concerns over sustainability that were perhaps an early form of 'greenwash' to hide their desire for direct control of such an essential medical resource. At ports abroad and at home unscrupulous traders were adulterating batches of cinchona with other bitter barks such as cherry and cassia. In addition, the demand for supplies of

[opposite] Peruvian bark tree, *Cinchona calisaya*, painted in the field by the intrepid botanical artist Marianne North (c. 1870).

The gathering and drying of cinchona bark in a Peruvian forest. C. Leplante, after Faguet (c. 1867).

A pressed herbarium specimen of *Cinchona pubescens* (formerly *C. succirubra*), collected by the Ruiz and Pavon expedition in Peru (1777–88) and now in Kew's collection.

cinchona bark to cure and control malaria increased, as empires turned to new territories in Africa with tropical diseases fatal to European colonisers.

A STEW OF SPECIES

The bark of the twenty-five species of cinchona varies greatly in quinine content, which is determined by genetics and environment. Some of the variation is explained by altitude but soil type, competition and soil microorganisms are probably important too. For Western botanists in the eighteenth and nineteenth centuries, the variability in quinine content in different species was baffling, although well-known to local peoples who well understood which environments and types of trees produced the best barks. The problem was amplified because scientists back in Europe were attempting to understand the botany

A French first-day cover from 1970, celebrating the 150th anniversary of the discovery of quinine by Caventou and Pelletier.

and pharmacology of cinchona through analysing dried bark samples landing at trade ports, and trying to cross-reference them to the few pressed plant specimens found in European botanical collections.

Cinchona trees are notoriously difficult to identify to species: they look very similar and are best told apart using subtle leaf characters. When different species grow nearby, they readily cross, leading to hybrid plants with mixed characters, further confusing identification. As with many useful plants, minor variants were often named as new species. Genuinely new species are still being found, with *Cinchona anderssonii* the most recent. It was found in the Bolivian Andes in 2013, recognised as a new species, and named in 2017 after Lennart Andersson (1948–2005), the Swedish botanist whose 1998 study finally brought order to the botanical naming of cinchona species.

In 1820, two French chemists, Joseph Bienaimé Caventou and Pierre-Joseph Pelletier, were the first to isolate two of the four main alkaloids from cinchona bark. They named them cinchonine, after the genus name, and quinine, after one of the trees common names, *quina*. Scientists now had a way to extract the cinchona alkaloids for measured dosage and could carry out precise clinical trials. Furthermore, if cinchona species growing in the wild could be accurately named and the quinine content of their barks measured, then those species richest in quinine could be selected for transplantation to colonies of a similar ecology and (crucially) availability of cheap labour for cultivation and harvesting.

Several earlier efforts had been made to transplant cinchona from the Andes to other parts of the world, including by the French in the 1730s. By 1813, British botanists working in India had proposed the transfer of cinchona trees to plantations there.

In 1851, Justus Karl Hasskarl travelled to Peru to obtain seeds at the behest of the Dutch Minister of Colonies. Disguised

MAUVE: AN ACCIDENTAL INVENTION

In the nineteenth century many chemists attempted to synthesise quinine, which would enable its manufacture in factories without the need to harvest bark. In 1856, eighteen-year-old William Henry Perkin attempted it using coal tar. He failed, but noticed that the resulting mixture was a deep purple. He was astute enough to experiment with his failed sludge to see if it would dye fast to cloth. It did, and he became the accidental inventor of the first synthetic aniline dye, mauveine. Before this, purple dye was only obtainable from expensive natural sources. The results led to an explosion in demand for purple clothing, previously only available to the highest classes in society, termed 'mauve madness' or, in *Punch* magazine, *'mauve measles'*.

as Dr José Carlos Muller to hide his true purpose from the authorities, he set about capturing cinchona. He obtained 500 live plants and sent them off, though all but 75 died before reaching the Dutch East Indies (Indonesia). The death of plants on the long journey was to plague many plant hunters and constrained the transfer of plants. It was later claimed that Hasskarl's assistant, a man named Henriquez, had poisoned the soil in the Wardian cases with arsenic to protect the traders' livelihoods. It was also alleged that South American collectors would also heat the seeds so that they would not germinate.

EMPIRE AND EXPEDITION

Historian Daniel Headrick has called cinchona, alongside other technologies such as the steamship and guns, a 'tool of empire'. During the second half of the nineteenth century, as European governments grappled for more territory and resources, tropical diseases were a major problem for colonisers. As well as bringing disease to uncontacted peoples around the globe, they were affected by newly encountered tropical diseases themselves. Unused to the challenges of extreme climate, harsh conditions and the novel diseases experienced during expeditions, death rates were disastrous. British troops sent to Sierra Leone between 1819 and 1836 suffered on average 483 deaths per 1,000 troops, and worse statistics abound: William Bolt's expedition to Mozambique lost 132 out of 152 Europeans and the Lander-Laird Niger expedition of 1832 lost 40 out of 49.

Peruvian bark had little effect against other tropical diseases such as yellow fever and dysentery, and until quinine became available in ready-made doses, the bark was not always considered an essential

[above] Mauve, the colour accidentally discovered during an attempt to synthesise quinine. Early 20th century postcard.

[opposite] David Livingstone (1813–1873), Scottish missionary and African explorer, being carried 'The Last Mile' as he lay dying. *The Life and Explorations of David Livingstone* (1887).

or reliable medicine for the traveller's kit. The Navy occasionally used quinine as a preventative from the late eighteenth century but its worth as a preventative and not just a treatment was not formally established until 1854, when Captain William Balfour Baikie (1824–1864) was commissioned to undertake an exploratory expedition up the Niger. Baikie was a physician and ensured that each of the Europeans under his command took a daily dose of quinine. All returned alive. The success of Baikie's expedition paved the way for more exploration and was followed by the 'scramble for Africa' between imperial nations.

David Livingstone (1813–1873), famous medical missionary and African explorer, also popularised the use of quinine as a preventative, having used it successfully for twenty years during missions to east and central Africa. His preferred method was to mix quinine with purgatives such as jalap and calomel, and he developed quinine in the form of pills which were eventually marketed as 'Livingstone's Rousers'. Livingstone's medical supplies, including his quinine, were stolen and he wrote in his diary *'I felt as if I had received the sentence of death'*. He died shortly after from a combination of malaria and dysentery.

THE HUNT FOR CINCHONA: THREE EXPLORERS

By 1850, the Indian government was spending the equivalent of £500,000 per year on imported cinchona, and cultivation became urgent for the imperial project. Kew Gardens, as the pre-eminent British botanical institute, played a major role in coordinating efforts. It built a glass forcing house for growing cinchona seedlings to send to India and provided both knowledge and experts to transfer the plants across the world. All that was needed was the tree itself.

In 1859, Clements Markham, working for the India Office, collaborated with Sir William Hooker, Director of Kew Gardens, to oversee an expedition to obtain cinchona. Markham was selected for his familiarity with the area, having previously travelled around Peru to write

THE WARDIAN CASE

In 1829, Dr Nathaniel Bagshaw Ward packed up a moth cocoon in a sealed jar and waited for it to emerge. During observations, he noticed ferns and grass had germinated in the soil at the base of the jar, so he moved the jar to a sunny window and watched them develop over the next three years. He had hit upon a discovery: within the stable environment of a sealed container, air and moisture could sustain plant life.

Ward had found the solution to the major problem of how to keep plants alive on long voyages that passed through different climates. Working with the London nursery owner George Loddiges, Ward sent the first Wardian case to and from Sydney, Australia, in 1833 and the plants' survival over a journey of several months proved the concept. The Wardian case, a miniature and portable greenhouse made of wood and glass, enabled the successful transfer of cinchona and thousands of other species across the globe until the 1960s, when airfreight took over.

its history, but he had little botanical knowledge. The political instability of South America nations, at a period shortly after most had become independent from Spain, meant his journeys would not go as planned. Markham targeted Peruvian cinchona trees. The quest was not an easy one; shortly after arriving in January 1860, he heard threats that anyone attempting to steal cinchona would have their feet chopped off. He collected plants from the Carabaya forests, before hearing of an order for his arrest. He quickly left for the coast with 450 plants packed into Wardian cases and sent them on to India via Kew by June of the same year.

Expert botanist Richard Spruce was a Yorkshire man who earned his credentials by publishing a list of his local plants by the age of 16, followed by a book just three years later. His skills attracted attention from William Hooker, who commissioned him to collect plants in the Amazon rainforest. He was already in South America at the time of Markham's recruitment, having arrived ten years earlier, in 1849. Spruce was sent to search for the most valued cinchona, the red-barked form, *cascarillo roja* of Ecuador, found near the mountains of Chimborazo. He was successful, gathering 637 plants and around 100,000 seeds, despite seemingly insurmountable problems of difficult terrain, poor health, and a mysterious and painful paralysis. He met Kew gardener Robert Cross and they set off down a river to transport the plants to the coast, where they could be loaded onto a boat. The river was flooded and the journey was treacherous. The plants were tossed about and almost capsized but were saved by their careful packing in Wardian cases. From the port of Guayaquil Robert Cross accompanied the plants onboard a steamer to Kew, arriving in October 1860. It seemed the British had secured cinchona for their plantations, but there is more to the story.

Charles Ledger was a British explorer who arrived in South America in the 1830s, settling on a career in farming. After collecting a herd of 619 alpaca, he drove them under treacherous conditions across the Andes mountains to Chile for transportation to Australia. He lost his money in the venture and turned to cinchona in 1861 after reading about the efforts to transplant it to India.

[above left] Sir Clements Robert Markham (1830–1916). Wellcome Collection.

[above centre] Richard Spruce (1817–1893).

[above right] Charles Ledger (1818–1905). Wellcome Collection.

[opposite] Wardian Case en-route to Sri Lanka in the late 19th century.

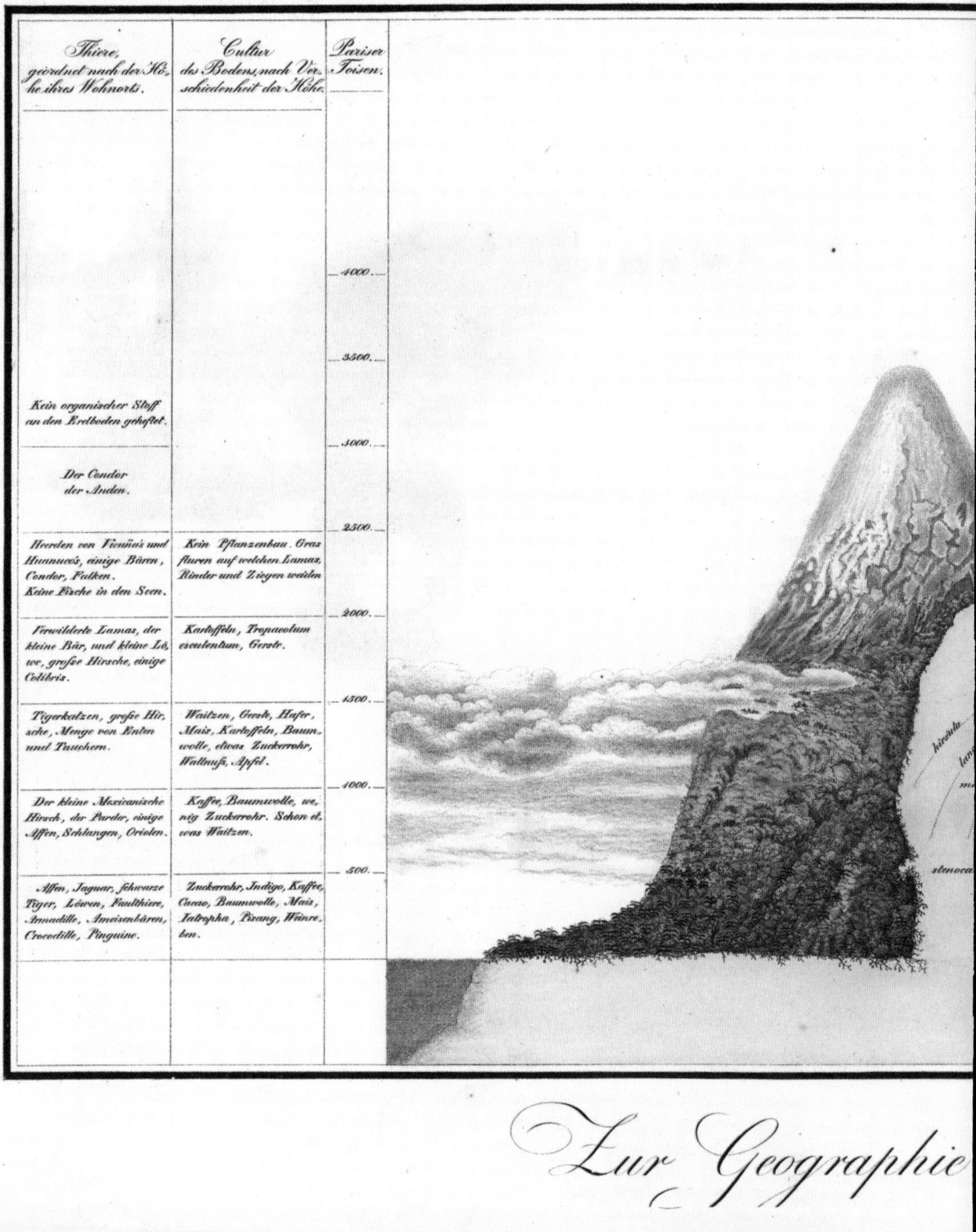

Thiere, geordnet nach der Höhe ihres Wohnorts.	Cultur des Bodens, nach Verschiedenheit der Höhe.	Pariser Toisen.
		4000.
		3500.
Kein organischer Stoff an den Erdboden geheftet.		
		3000.
Der Condor der Anden.		
		2500.
Heerden von Vicuña's und Huanuco's, einige Bären, Condor, Falken. Keine Fische in den Seen.	Kein Pflanzenbau. Grasfluren auf welchen Lamas, Rinder und Ziegen weiden	
		2000.
Verwilderte Lamas, der kleine Bär, und kleine Löwe, grosse Hirsche, einige Colibris.	Kartoffeln, Tropaeolum esculentum, Gerste.	
		1500.
Tigerkatzen, grosse Hirsche, Menge von Enten und Tauchern.	Waitzen, Gerste, Hafer, Mais, Kartoffeln, Baumwolle, etwas Zuckerrohr, Wallnuss, Apfel.	
		1000.
Der kleine Mexicanische Hirsch, der Parder, einige Affen, Schlangen, Oriolen.	Kaffee, Baumwolle, wenig Zuckerrohr. Schon etwas Waitzen.	
		500.
Affen, Jaguar, schwarze Tiger, Löwen, Faulthiere, Armadille, Ameisenbären, Crocodille, Pinguine.	Zuckerrohr, Indigo, Kaffee, Cacao, Baumwolle, Mais, Iatropha, Pisang, Weinreben.	

Zur Geographie

This illustration was based on Alexander von Humboldt's famous Naturgemälde (nature painting) of 1807, an early infographic that showed variation in the vegetation of the Andes.

Altitudinal map of cinchona in the Andes, from *Versuch einer Monographie der China* by Heinrich Von Bergen (1826).

Rheinl. Fuſs.	Höhen meſsungen in verschiedenen Welttheilen. In Rheinländischen Fuſs.	Höhe der untern Grenze des ewigen Schnees, nach Verschiedenheit der geographischen Breite. In Rheinländischen Fuſs.
27000.	Höhe der kleinsten Wolken (Schäfchen) Höhe des Dhawalageri, Gipfel des Himalaya, nach Marsden. 25690.	
24000.		
21000.	Gipfel des Chimborazo, nach Humboldt. 20852. Gipfel des Cotopaxi, nach Bouguer. 18332.	
18000.	Gipfel des Popocatepel, nach Humboldt. 17164. Pic von Orizava, nach Humboldt. 16904. Montblanc, nach Sauſsure 15215.	Unterm Aequator und 3 Grad N und S, nach Humboldt. 15300.
15000.	Spitze des Ortles, nach Gebhard. 14920. Finsteraarhorn, nach Tralles. 13898. Iungfrau, nach Tralles 13320.	Unter 20 Grad N nach Humboldt. 14660.
12000.	Pico de Teyde, nach Cordier. 11805. Aetna, nach Sauſsure. 10638.	Unter 35 Grad N, nach Humboldt. 11180. Unter 40 Grad N, nach Humboldt. 9930.
9000.	Paſs über den Groſsen Bernhard nach Sauſsure. 7738.	Unter 45 Grad N, nach Wahlenberg. 8500.
6000.	Vesuv, nach Shukburg. 3819. Brocken, nach Lindenau. 3760. Hecla, nach Povelsen. 3229.	
3000.		

nitida
micrantha
Mutisii
Condaminea
Pavonii
cordifolia
rotundifolia
oblongifolia
dichotoma
grandiflora
magnifolia
Humboldtiana
pubescens
purpurea
rosea
caduciflora

der Cinchonen.

Ledger had an advantage: his friend and assistant of over twenty years, Manuel Incra Mamani, a native Bolivian with detailed knowledge of his environment. Mamani had shown his expertise in cinchona quality during previous expeditions and his local knowledge was essential to Ledger, who commissioned him to obtain seeds of the richest types of cinchona. When asked by Ledger if an area of cinchona trees they passed was valuable, he replied '*No Senor, the trees here about do not see the snow-capped mountains*' (the Andes). After locating promising trees, he missed the flowering season but finally, after four years, Ledger took delivery of forty pounds of seeds in 1865. Unfortunately, his luck had not improved. When he tried to sell the seeds to Kew, William Hooker had just died, Kew had Spruce's seeds growing in India, and, in the absence of other experts, the seeds were turned away. At his wit's end, Ledger cast around for another buyer and was advised by John Eliot Howard, the London pharmaceutical manufacturer and cinchona expert, to try the Dutch. He sold them a pound of seeds for £20 (just over £1,000 in today's money), and a small amount to a private planter in India. Tragically, Mamani was imprisoned and beaten for his part in smuggling seeds and died shortly after – but the seeds, though late to the story, were to have significant impact on cinchona cultivation in Asia.

PLANTATION YEARS

The quest for quinine was not over. The plants were sailing towards their respective plantations but the value of their alkaloid content could not be assessed until the trees became established, which would take fifteen years. Selecting which trees to plant was a gamble and being the first to obtain seeds did not guarantee success. The cinchonas collected by Hasskarl, Markham, Spruce and Ledger now had to reveal their worth. It was a waiting game.

Hasskarl had returned in 1851 to Dutch-controlled Java, taking seventy-five plants to the Cibodas plantation. He named them *Cinchona pahudiana* after the Dutch minister of colonies who had initiated the expedition. They turned out [right] Single-dose quinine sachet distributed at Indian Post Offices.

[opposite] *Cinchona pubescens* saplings growing at Kew.

to have very low levels of quinine in their bark, approximately 0.2 per cent.

While the Dutch were discovering their trees were not as valuable as they hoped, Markham's 450 seedlings were heading to government plantations in India. However, after a disastrous journey of delays, becalming, and harsh weather conditions, none survived planting. Spruce and Cross's plants of *Cinchona succirubra* and *C. officinalis* fared better. Their barks had a higher alkaloid content, between one and five per cent quinine, as well as other effective quinoline alkaloids (cinchonine, cinchonidine and quinidine), which meant the barks were better suited to producing antimalarial drugs.

Ledger and Mamani's seeds, rejected by Kew and now in the hands of the Dutch, were named *Cinchona ledgeriana* in honour of the trader who had brought them to the attention of the Dutch. The plants grown from these turned out to thrive in Indonesian conditions and had remarkably high amounts of pure quinine alkaloid, between 5 and 13 per cent. The Dutch carefully crossbred the highest-producing trees.

Meanwhile, in British Indian plantations, some of Ledger's seeds sold to private planters and collectors had ended up in the government plantations, but they were not suited to the environment of India and did not thrive. The British turned to creating antimalarials from the mixture of cinchona alkaloids found in Indian plantation barks, which medical trials had shown in many cases to be as effective as quinine. This enabled the provision of medicines at cost price for use within their army and (in ambition if not in practice) for distribution to the workers of India.

The Dutch and British empires had two very different motivations for cultivating cinchona. The Dutch grew *Cinchona ledgeriana* in Java for its high quinine content, primarily for the export market. The British in India deliberately grew a range of species and hybrids best adapted for its climate, and rich in all four quinoline alkaloids, primarily for local consumption. By 1883, cinchona barks from south and South East Asia dominated the market, ensuring an in-house supply for empire and wiping out the South American trade.

IMPERIAL MOTIVES

'. . . *few greater blessings could be conferred on the human race than the naturalisation of these trees in India and other congenial regions, so as to render the supply more certain, cheaper and more abundant.*'
(Clements Markham)

Pl. XI
CINCHONA OFFICINALIS,

VIII

[above] View of the Salak Volcano, Java, from Buitenzorg by Marianne North (1880).

opposite:

[top left] *Cinchona calisaya* (formerly *C. pahudiana*) from *Illustrations of the Nueva Quinologia of Pavon* by John Eliot Howard, illustrated by William Fitch (1862).

[top right] *Cinchona officinalis*, from *The Quinology of the East Indian Plantations* by John Eliot Howard, illustrated by William Fitch (1869).

[bottom left] *Cinchona calisaya* var. *ledgeriana*, from *The Quinology of the East Indian Plantations* by John Eliot Howard, illustrated by William Fitch (1869).

[bottom right] *Cinchona pubescens* (formerly *C. succirubra*), from *Illustrations of the Nueva Quinologia of Pavon* by John Eliot Howard, illustrated by William Fitch (1862).

KEW'S CINCHONA COLLECTIONS

The largest collection of cinchona barks and pressed plants (herbarium specimens) in the world resides at Kew. Dating from the late eighteenth century to the present day, there are around 1,000 barks and artefacts in the Economic Botany Collection and nearly 1,000 pressed plants. Kew's Library, Art and Archives also hold a wide range of original sources, ranging from exquisite volumes of botanical illustrations to the 'Miscellaneous Reports', the bundled archives of correspondence and reports about cinchona cultivation in nineteenth century British colonies such as India and Sri Lanka.

About half of Kew's barks were collected by John Eliot Howard (1807–1883). He was born into a family of pharmaceutical manufacturers and from an early age established a reputation as a dedicated chemist, botanist and 'quinologist': an expert in cinchona and quinine. The family business, Howards and Sons, was based in Stratford, the site of the 2012 Olympics in east London, and led the quinine industry from the mid nineteenth century until the early twentieth century.

Howard collected cinchona bark at the London docks and trading rooms, and analysed them chemically and microscopically, publishing prolifically on the identification of cinchona from barks (p. 10 and book endpapers). As well as supporting his family business interests, Howard advised the government on the introduction of cinchona trees to India.

The barks in the collection represent Howard's lifelong work in understanding the botany and chemistry of cinchona, and represent nineteenth-century Andean cloud forests that are now sadly depleted. These barks are still actively used in scientific research today, providing samples for chemical and genetic analysis by researchers at Kew, the University of Copenhagen and the National Herbarium of Bolivia. This is casting new light on the combination of evolution and environment that controls quinine production in the cinchona tree.

The nineteenth century was a period of plant exchange on an unprecedented scale. As the quotation on p. 53 suggests, empire-builders such as Markham gave little thought the principles of national sovereignty, informed consent and benefit-sharing that govern plant transfers today. Instead, he vividly describes his efforts to escape the officers of local government in his journey to the Peruvian coast. Those involved in the cinchona project justified their actions on humanitarian grounds – and considerable effort was given to making quinine as widely available as possible in India, through sale at the lowest possible price through post offices. The health of India's native population was important to the British occupiers because a healthy workforce would be more productive. The motivations and real impact on public health of the cinchona project continue to be debated by historians. What is clear is that it took place under the auspices of repressive empires and was driven by the demands of trade and colonialism. In this context plants were moved around the world with a careless regard for the rights of peoples or nations, or for the effects on the environment of their cultivation.

[above] John Eliot Howard (1807–1883). Howard collected and analysed many of Kew's cinchona specimens.

[opposite] Economic Botany Collection no. 52935. A bundle of cinchona bark with label detailing the chemical content analysed by John Eliot Howard.

DECLINE OF QUININE AS A MEDICINE

In 1942 Japan took control of Indonesian cinchona plantations, which had been providing bark for American quinine production. With America now at war with Japan, sourcing quinine or any alternatives would be vital tools for keeping troops alive in tropical climates.

US chemists tested a German antimalarial synthetic compound that had been handed over by French authorities in 1943. It had been created in 1934 by Johann 'Hans' Andersag, a chemist for the German pharmaceutical company Bayer. It had been patented but put aside as too toxic for human use. Clinical trials showed it to be highly effective, and renamed as chloroquine it was widely used after 1946. Quinine's seat as the major antimalarial was toppled. While the natural alkaloids continued to be used, a new era of synthetics was taking over from drugs sourced from plants. The shift away from quinine was also driven by its toxicity, and recognition from the 1890s that malarial parasites were becoming sporadically resistant to quinine, making it increasingly ineffectual in some regions.

Quinine is still used today in some cases of lupus and rheumatoid arthritis but is only recommended for malaria by the World Health Organization as a last resort in patients with a drug-resistant form. Quinine has other medical uses and is sometimes prescribed to alleviate leg cramps. Some sufferers swear by a glass of tonic water before bedtime, although the amount of quinine is probably too small to have an effect.

[top] Clearing virgin forest in Java, probably late 19th century. Wellcome Collection.

[centre] Planting cinchona trees, Java. Wellcome Collection.

[bottom] Harvesting and processing cinchona bark on a Java plantation.

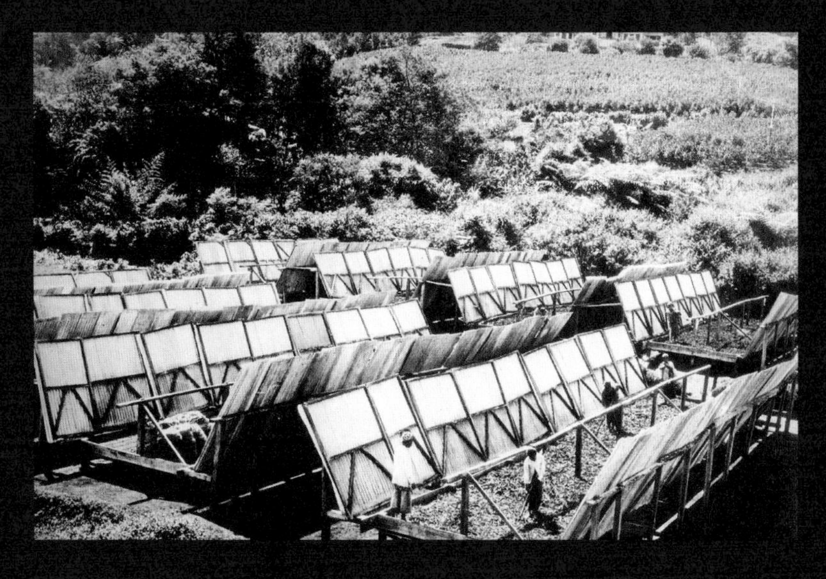

[top] Drying cinchona bark in the sun, Java. Wellcome Collection.

[centre] Processing cinchona bark in a Dutch–Javanese quinine factory. Wellcome Collection.

[bottom] Interior of quinine extraction factory, Mangpoo, India. Wellcome Collection.

SOLUBLE
ESSENCE
IMPERIAL QUININE TONIC
RECIPE.
TRADE MARK REGISTERED
HEART BRAND
DUCKWORTH & Co.
Old Trafford Essence Distillery,
Chester Road, - MANCHESTER.
DUCKWORTH & CO.

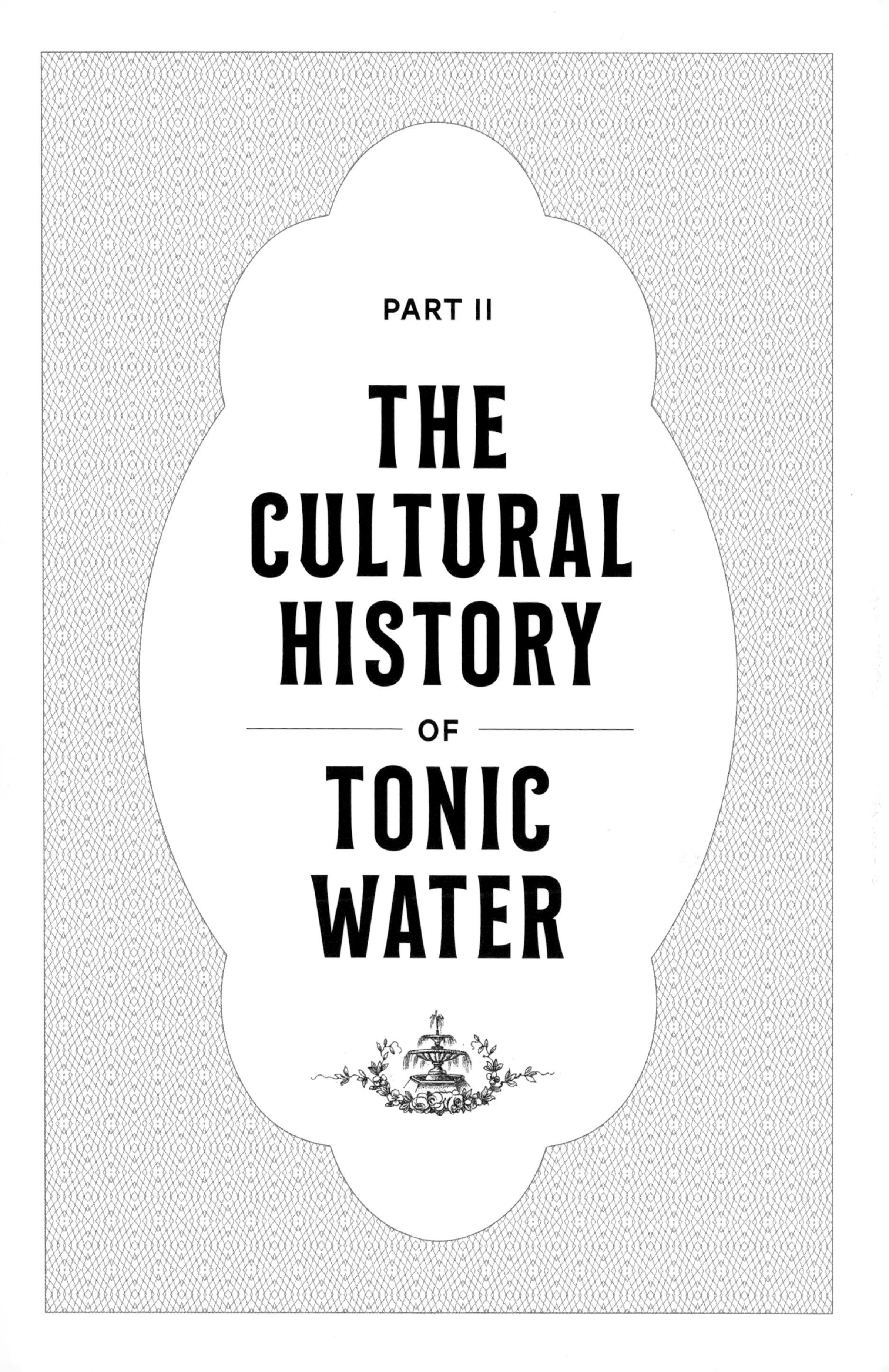

PART II

THE CULTURAL HISTORY OF TONIC WATER

HAYWARD
TYLER
& CO.
LONDON
HARE

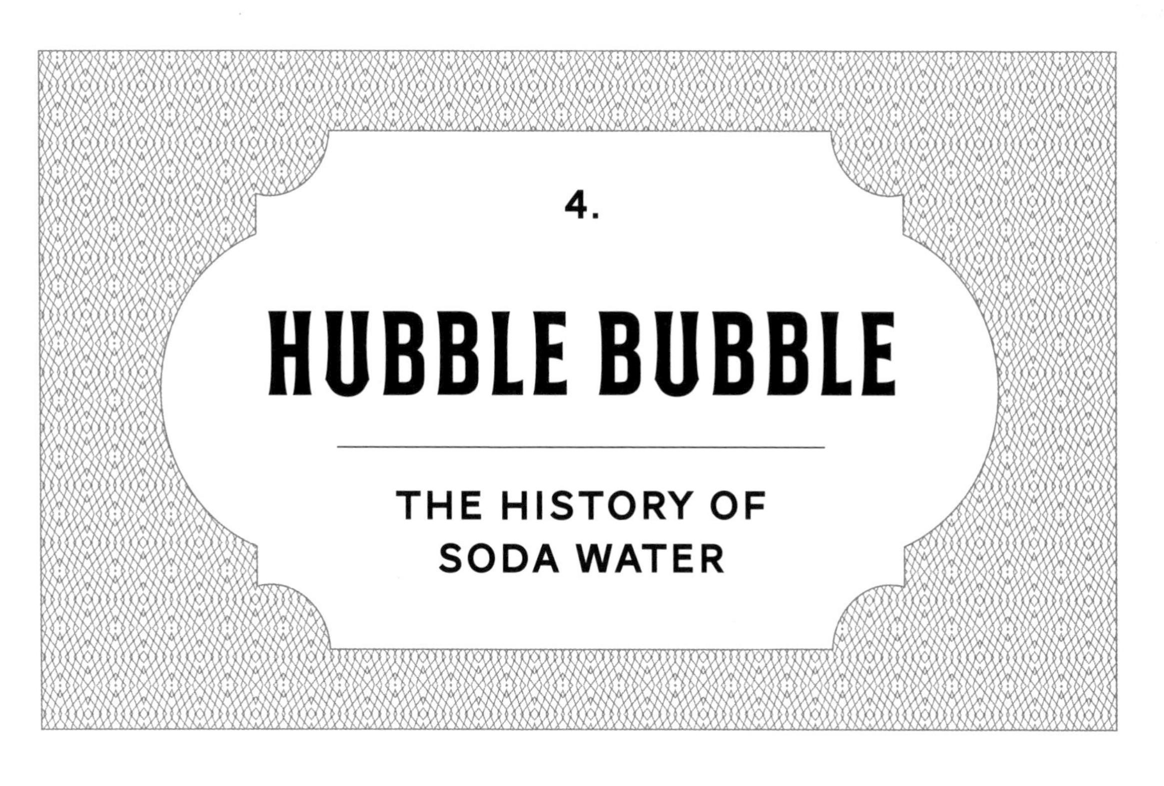

4. HUBBLE BUBBLE

THE HISTORY OF SODA WATER

Soda, in the sense of a plain fizzy water, rather than the American usage for flavoured sparkling soft drinks, no longer enjoys widespread popularity. Though you will find it in all pubs, it is relegated to the plastic siphon as carbonated tap water or bottled in small servings to mix with cordial, as an occasional splash to liven up spirits. However, soda, or aerated water, was once a sensation valued not just as a refreshingly bubbly drink but for its medicinal properties. The invention of soda water combines two histories: the belief in the curative powers of mineral water, and the eighteenth-century discovery of carbon dioxide and its scintillating properties.

[opposite] Hayward Tyler & Co. Soda water machinery, *The Chemist and Druggist* (1880). Wellcome Collection.

[previous page] Imperial Quinine Tonic (c. 1900). Wellcome Collection.

A QUICK DIP: THE HISTORY OF MEDICINAL MINERAL WATERS

From ancient Egyptian beauty regimes using water vapour, to the medicinal hydrotherapy prescribed by the ancient Greeks: drinking, washing, bathing and steaming with hot or cold, fresh, sea, and mineral water has a long history for health and pleasure. The most celebrated of ancient Greek medical texts, the Hippocratic corpus, dating to the fourth and fifth centuries BC, assigns different mineral waters distinct properties for

The thermal baths at Bath, Somerset, from *The English Spy* by Robert Cruickshank (1825). British Library.

different maladies. The popularity of bathing in Europe rose with the ascendance of the Roman Empire and in ancient cities bathing became part of everyday life, mixing pleasure and wellbeing.

The Belgian town of Spa, home to famous mineral springs, has lent its name to water therapy. The popularity of spas and mineral bathing had a resurgence in the eighteenth and nineteenth centuries, not only for cure and convalescence, but for socialising among the upper classes. The Somerset town of Bath is famous for its spa, providing a backdrop for many novels of the period including Jane Austen's *Persuasion*. Mineral baths were a destination for mixing of the sexes, where romantic matches could be made alongside taking the waters for health.

Healthy springs and spas, some with naturally sparkling water, contrasted with stagnant malarial waters. They led to an interesting question: what made some waters healthy and others not? During the eighteenth and nineteenth centuries, scientists experimented with mineral waters, attempting to discover their distinctive healing properties. This flurry of research gives the context for the invention of a mass-produced mineral water.

JOSEPH PRIESTLEY AND THE INVENTION OF AERATED WATER

Developing sparkling water from the natural spring to the factory depended on the efforts of a long chain of scientists who discovered the nature of gases and mineral waters, and experimenters who combined the two. The central invention involved the dissolving of carbon dioxide (CO_2) under pressure in water, which readily absorbed it, to give it the '*bubbling scintillation*' as one writer described.

In 1755 Joseph Black (1728–1799) showed that in dissolving limestone with acid, a gas (now called carbon dioxide) was produced. This '*elastic fluid*' was named '*fixed air*' and would snuff out candle flames. Ten years later, William Brownrigg (1711–1800) connected this same fixed air with mineral waters, presenting to the Royal Society *An*

Joseph Priestley (1733–1804), inventor of aerated water (1782). Wellcome Collection.

Experimental Enquiry into the Mineral Elastic Spirit of Air, contained in Spa Water. Brownrigg described a trip to the eponymous Belgian town of Spa to conduct investigations on its waters. In an ingeniously simple experiment, he took a glass bottle of the mineral water and tied a collecting balloon made from a calf's bladder over the mouth. Once the water was heated, the balloon expanded, showing the release of gas from the water, which he then showed to be Black's elastic air. The spent waters became cloudy, losing their flavour and freshness and showing that the air was responsible for its health-giving properties. Brownrigg also noted that these gases were produced during fermentation, another topic with which the inventor of carbonated water, Joseph Priestley, would experiment.

Joseph Priestley (1733–1804) was a Leeds-born dissenting minister with a remarkably wide range of interests. He

Two men take soda water for a hangover.
Postcard (1913). Wellcome Collection.

DAY MORNING

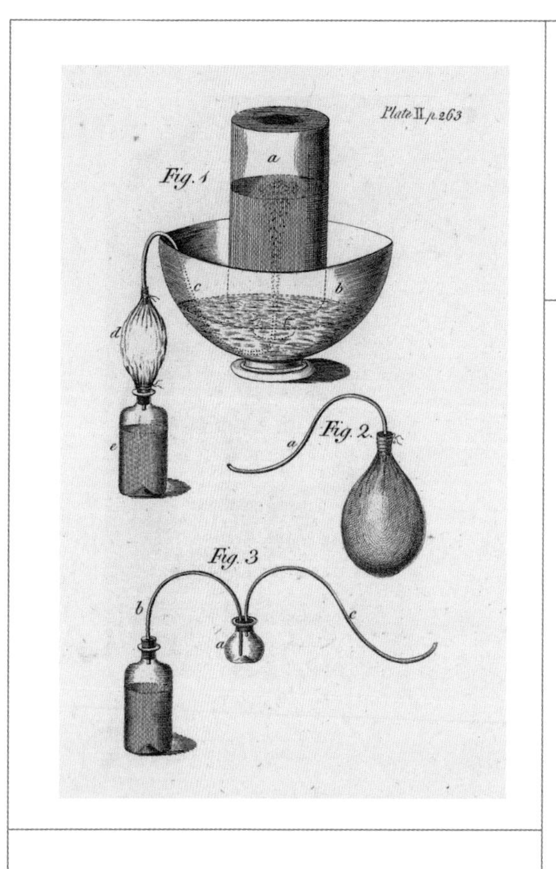

[left] Joseph Priestley's apparatus, from *Directions for Impregnating Water with Fixed Air* (1772). Wellcome Collection.

[opposite] John Mervin Nooth's 1774 interlocking glass vials were an improvement on Priestley's design for aerating water. Science Museum/Wellcome Collection.

published extensively on electricity, light, theology, natural philosophy, education and chemistry. He is best remembered for his work with gaseous elements, including the discovery of oxygen and the anaesthetists' nitrous oxide.

Priestley was fascinated with gaseous elements and experimented with Black's fixed air he found arising from the fermenting vats in a neighbouring brewery. Priestley floated a saucer of water in a beer vat overnight, and the next day noticed that the water had a *'very sensible and pleasant impregnation; and it was with peculiar satisfaction that I first drank this water, which I believe was the first of its kind that had ever been tasted by man'*. However, he was soon banished from the brewery by an infuriated manager after accidentally ruining a keg of beer by knocking his experimental chemicals into it. Not to be deterred, he, like many practical-minded early scientists, continued his investigations at home using make-do equipment of beer glasses and earthen washing vats. In 1772 he published his *Directions for Impregnating Water with Fixed Air*, which illustrated the method for creating aerated waters: a glass bottle containing chalk dissolved in acid released carbon dioxide. The bottle was connected via a leather tube to an inverted and partially submerged jar of water. As the gas passed through, it dissolved into it, making the water fizzy. Priestley also suggested, perhaps remembering the disgruntled brewery owner, that these waters could be used to revitalise flat beer, a foreshadowing of the invention of keg beer.

AERATED WATER FOR SCURVY

Priestley's invention of aerated waters coincided with the theories of naval surgeon David MacBride. MacBride had suggested in 1764 that fixed air could be used to prevent putrefaction of the flesh. He believed that the use of Priestley's novel impregnated waters could thus treat disorders such as putrid throat, leg ulcers and, vitally, scurvy. It was this latter complaint that received attention from the British Admiralty, being a scourge among sailors on increasingly long naval voyages. Scurvy causes weakness, joint pain, bulging eyes, scaly dry skin and spongy gums.

MacBride's influence resulted in the Admiralty adopting a plan to distil fresh water from seawater and carbonate it during voyages. With the approval of the Royal College of Physicians, two ships, HMS *Resolution* and *Discovery* – the latter on which James Cook was about to commence his second voyage to the Pacific – were kitted out with equipment designed by Priestley. Sadly, aerated waters are ineffective at treating scurvy, and the true cure, fruits containing vitamin C, took another forty years to be adopted. Scottish Doctor James Lind had demonstrated in 1753 the curative effects of citrus for scurvy in one of the earliest recorded medical trials, but had to wait until the 1790s before his suggestions were literally taken on-board in the form of casks of lemon juice. This later changed to lime juice, furnishing British sailors with the nickname 'Limeys'.

Priestley never made any financial gains for his discovery of aerated water but he was presented with the Royal Society's Copley Medal in 1773, for his contribution towards treatment of scurvy. It wasn't until over 100 years later, in 1874, that a monument to his contribution to the medicinal and soft drink industry was unveiled in Birmingham with a speech celebrating his '*service to . . . thirsty souls, which those parched throats and hot heads are cooled with morning draughts . . . cannot too gratefully acknowledge*'.

IMPROVING APPARATUS

Priestley's design for impregnating waters with fixed air went through cycles of design improvements, most notably in 1774 by John Mervin Nooth (1712–1828). He designed a more productive method of manufacture using an airtight method of interlocking glass vials with a one-way valve structure. He called it the gazogene. It could be sold ready-made and was compact enough to be kept on the sideboard at home. Unfortunately, part of the valve was liable to jam, and the build-up of pressure could result in explosions. In 1846 it was adapted by Robert Liston and Peter Squire as a gas-producing inhaler for general anaesthesia. Successive inventors improved on Nooth's apparatus, using different chemicals to replicate the

mineral waters from various spas, which could then be prescribed by physicians for different ailments, and be made available over the pharmacy counter.

SODA WATER AND THE SCHWEPPES EMPIRE

By the turn of the nineteenth century, aerated waters were popular as a panacea for general health, convalescence and a range of disorders ranging from gout to rheumatism and digestive disorders. Medical journals and newspapers advertised soda fountains for pharmacies or the regular home user, and bottled soda waters ready for use. The earliest glass bottle that could withstand the pressure of the aerated waters was the returnable (two shillings for every dozen) egg-shaped bottle, also known as the Hamilton. It was designed in 1809 by William

The curative effects of soda water. Image by George Cruickshank, (c. 1840). Wellcome Collection.

Hamilton in association with the Dublin firm Austin Thwaites, who had the first patent for soda water.

Up to the 1760s, aerated waters weren't quite the same as the soda water we know today; they lacked the addition of sodium bicarbonate that makes a true soda water. The addition of the salty substance was suggested in 1767 by a Norfolk apothecary named Richard Bewley, who found it helped the absorption of fixed air. He sold it under the name of 'mephitic julep'. It wasn't until 1799, with Thwaites patenting it under the name soda water, and the slightly later use by Schweppes, that the term became widely used.

The name of the man who made soda water into a truly popular product still appears on many bottles. Jacob Schweppe was born in 1740 to an agricultural family in what is today Germany. Schweppe showed an adeptness with his hands from an early age and his creative flair saw his rise from workshop assistant to silversmith apprentice. He then became a jeweller, and eventually settled in Geneva. Influenced by Priestley's work on aerated waters, and handy at practical problems, he tweaked the original designs aiming

[left] Early Schweppes advertisement. From *Paterson's Guide to Edinburgh* (1883). British Library.

[below] Soda water machinery was commonly advertised throughout the 19th century. *The Chemist and Druggist* 1892. Wellcome Collection.

to find the perfect machine. Schweppe preferred tinkering with machinery to business matters, and the excess waters he produced were given freely to the poorest citizens of Geneva. Demand rose as he gained fame and by 1783 he had begun his commercial enterprise.

Schweppe worked on his aerated waters for ten years before partnering with Nicolas and Jacques Paul – a father and son team who had created a similar machine – and their friend Henry Gosse. A company was established in 1790 as Schweppe, Paul & Gosse, but ran into trouble, with disagreements on who designed the equipment. The Pauls were engineers and instrument makers who had been put in charge of the improvement of the Geneva water pump. They helped to improve the Schweppes machine and created a system known as the Geneva apparatus, in which the mechanics were enclosed and false parts were added to avoid replication by competitors. Using this equipment, they replicated the mineral waters of various eminent spas, including the seltzer and spa waters of Pyrmont, Germany. As they grew more popular, they looked further abroad to England.

The first British factory was opened in London at 141 Drury Lane, but business was slow, competition high and prospects did not look good. Schweppe had come to England on the condition that he would leave if business did not improve, but he decided to stay and the partnership with Paul and Gosse was dissolved. Now unfettered by the internal rumblings of his previous partnership, Schweppe sold stoneware bottles of aerated waters, and soon attracted supporters including Erasmus Darwin (grandfather of Charles), which helped establish the credibility of his product. Schweppe continued from strength to strength, defeating his competition by perfecting his machinery and exercising canny business sense, such as giving

The glass fountain in the Crystal Palace, from *Dickinson's Comprehensive Pictures of the Great Exhibition of 1851*. Smithsonian Libraries

Marble Draught Stand soda fountain, from *Harper's New Monthly Magazine* (1872). Wikimedia Commons.

discounts for large orders. Jacob Schweppe retired and sold his company in 1798, and his successors built the brand up to become one of the most successful in the world. In 1851 Schweppes were the official provider of refreshments at the Great Exhibition, in the Crystal Palace. The great hall contained a fountain made of four tons of pink glass, which can to this day be found depicted on Schweppes labels and impressed onto each bottle.

PHARMACY FOUNTAINS

The popularity of the health-giving properties of soda water was not lost on enterprising pharmacists, and soda machines and fountains became a popular fixture in pharmacies in the US and Europe from the mid-nineteenth century. A range of models and patents were available: from discreet undercounter designs to proud countertop versions bedecked with columns and statuary. Soda water, on a blurred spectrum of health and pleasure, spurred heated debate in the pages of the pharmacy journal *The Chemist and Druggist*. On the one hand it recommended cordial syrups to add to aerated waters, and noted that one pharmacist, in mind of the famous adage *In vino veritas* (In wine lies truth), wrote on his soda dispenser *In soda sanitas* (In soda lies health). On the other hand, some worried that reliance on the soda fountain was driving pharmacy away from the technical side. An 1877 article bemoaned that *'the soda cocktail and cigar departments have received more attention in cost of furniture and convenience of arrangement than the department for compounding prescriptions'*. There were also worrying reports of the occasional deadly explosion from the build-up of pressure. Some complained that the British should be *'above the soda fountain business the mainstay of the American pharmacist'*, and that they were *'men of education, not message-boys and soda fountain attendants'*. However, it provided an important monetary top-up

and soda water could be found in British pharmacies well into the twentieth century.

Today the popularity of plain and flavoured soda water still brings in significant sales for the global soft drinks industry, which began with a group of inventors whose endeavour put a little sparkle in our drinks, to whom we should raise a glass.

[above left] Lemon syrup cordial for adding to soda water. *Wellcome, Burroughs & Co Catalogue* (1892). Wellcome Collection.

[above right] Advertisements for a range of flavouring syrups for soda water are commonly found in newspapers and journals throughout the nineteenth century. *The Chemist and Druggist* (1892). Wellcome Collection.

[below] Soda water siphons designed by Messrs. Mayo & Co. *Pharmaceutical Journal and Transactions* (1845).

ORANGE
QUININE
WINE
Prepared according to the
BRITISH PHARMACOPŒIA 1898.
DOSE. One wineglassful 2 or 3 times a day. Each ounce contains one grain of pure hydrochloride of quinine.

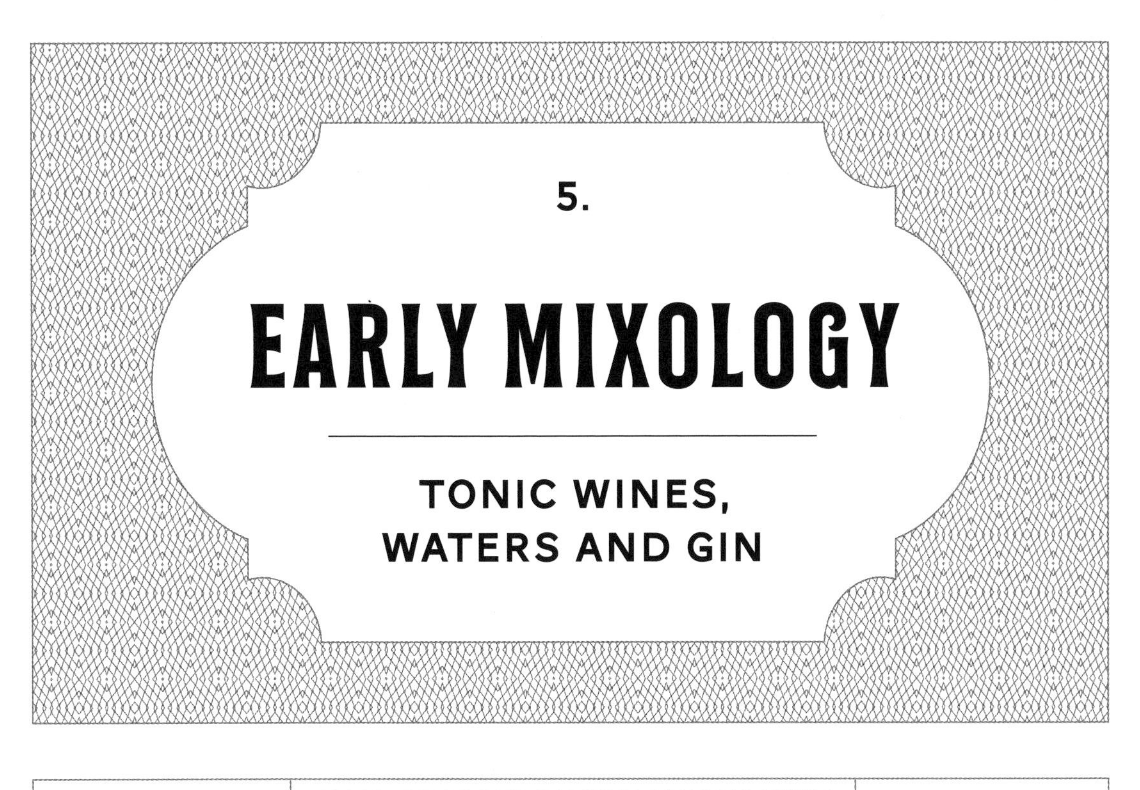

5.

EARLY MIXOLOGY

TONIC WINES, WATERS AND GIN

By the early eighteenth century, the quinine-containing barks of the cinchona tree were widely used in Europe and its colonies as a treatment for malaria. However, cinchona bark was also used as a treatment for many other illnesses, including dysentery, sore throat, toothache, smallpox, tremors and (externally) for baldness. Its function in treating these was usually as a tonic, for which purpose cinchona bark was highly regarded. Tonics were a significant category of medicine for much of the last 300 years, and remain important in Ayurvedic and Chinese systems of herbal medicine today. Tonics were considered to have little effect on the healthy, but to be vital in restoring the muscle tone and strength of the weakened body.

Cinchona bark, or from the 1820s, purified alkaloids, could thus be consumed as an antimalarial, or to treat other illnesses, or more generally as a tonic. However, the powdered texture of bark and the bitterness of the quinine within meant that it was not pleasant for the patient to consume. In response, cinchona bark or quinine was typically consumed in an alcoholic blend, often as medicinal tonic wines or mixed with rum or brandy.

TONIC WINES AND EARLY ALCOHOLIC MIXES

During the nineteenth century, sales of patent and proprietary medicines experienced a surge in popularity, and advertisements in newspapers, handbills and billboards publicised

[opposite] Orange Quinine Wine (1898). Wellcome Collection.

Hall's Coca Wine. Stephen Smith & Co. (c. 1899). Wellcome Collection.

these adveefashionable panaceas using beautiful fonts, pictures, catchy slogans and physician endorsements. These medications listed cures for a plethora of vague conditions including debilitated nerves, weakness of the stomach, torpid livers and female complaints. They appealed to all classes: from the rich, who invested in miracle cures and tonics, to the less well-off, who, in this pre-National Health Service era, could not afford visits to the doctor.

Tonics, bitters and fortified wines were popular forms of these medicines and were promoted as preventatives, cures and convalescent remedies. Tonic wines were often criticised, with some in the medical establishment labelling their claims as quackery and warning they could lead into alcoholism.

Tonic wines usually contained a fortified wine or spirit base, such as

An example of a tonic wine. From *Weeks & Potter Wholesale Druggist Catalogue* (1890).

sherry or port, with added enhancements including meat, malt, chocolate, coffee, quinine and even cocaine. Some famous brands that originated from these medicinal drinks still exist today, including Buckfast Tonic Wine and Coca-Cola. The latter originally included among its ingredients cocaine-containing coca leaf and caffeine-containing kola nut as a stimulant 'pick-me-ups'. Coca-Cola now uses coca leaves from which the cocaine has been removed, and the wine has been replaced by soda water and sugar syrup.

The new wonder-drug quinine with its established use against fever (a common side effect of many illnesses) and many other attributed cures from colds to hysteria, was a popular cure-all ingredient. A variety of quinine tonic wines emerged, including the sumptuous-sounding orange quinine wine made with

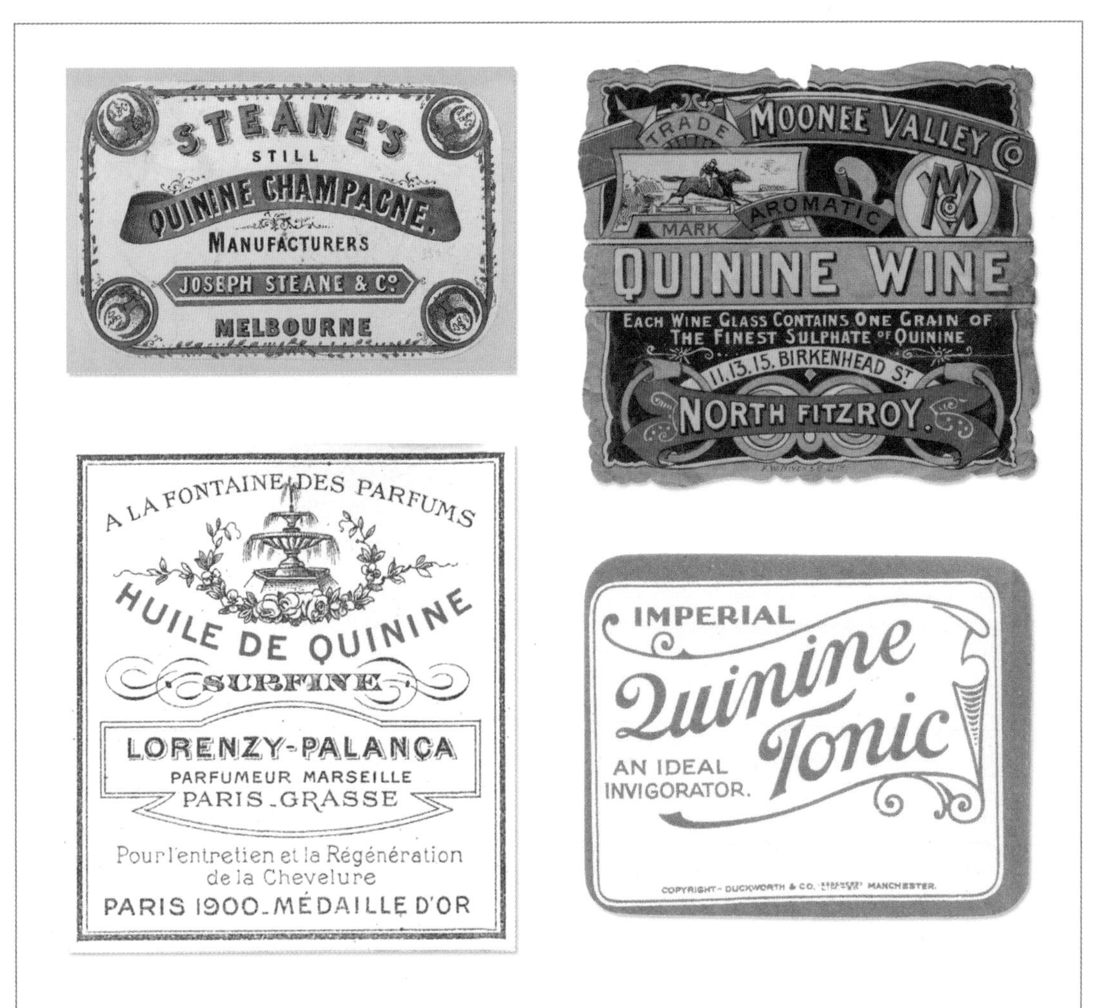

real Seville oranges and the somewhat less appetising beef and iron wine with added quinine.

APERITIFS AND DIGESTIFS

There was a fine line between drinks with health-giving tonic properties and those drunk for pleasure. What we consider aperitifs and digestifs, pre- and post-dinner drinks respectively, had their roots in bitter tonics containing a mixture of medicinal barks, roots and herbs. These tonics were used to stimulate digestion but are now considered more of an enjoyable tipple, just like the still popular pre-meal gin and tonic. Two aperitifs still containing bitter cinchona bark are the Italian China-Rossi and the French wine-based St. Raphael, invented in 1830 by Dr Adémar Juppet. The brand legend tells how he almost lost his sight while working late into the night creating his elixir. He prayed to the archangel Raphael to help him continue, was cured and

[top left] Steane's still Quinine Champagne (1871). State Library Victoria.

[top right] Quinine wine label. The Moonee Valley is a suburb of Melbourne, Australia. ABCR Auctions.

[bottom left] Quinine perfume based on the lilac scent of cinchona flowers was a popular fragrance in the 19th century.

[bottom right] Quinine tonics remained popular well into the 20th century. This label is dated to 1940.

CHINA-ROSSI

ELIXIR-TONICO-RICOSTITUENTE

FRA FRA

FRAT.LLI ROSSI-ASIAGO

China-Rossi for health and vitality.
Wellcome Collection.

BECKETT'S

TEMPERANCE DRINKS.

Highest Award, International Exhibition, London, 1885.

Can be USED WITH EITHER PLAIN or AERATED WATER

EXCELLENT FOR GAZOGENES AND SYPHONS.

Beckett's Lime Fruit Syrup.
Beckett's Raspberry Syrup.
Beckett's Black Currant Syrup.
Beckett's Lemon Syrup.
Beckett's Ginger Lemon Syrup.
Beckett's Gingerette Cordial.
Beckett's Winterine.
Beckett's Honey Liqueurs, &c., &c.

TONIC DRINKS:—

Beckett's Syrup of Orange and Quinine.
Beckett's Lime and Quinine.
Beckett's Syrup of Hops.

SPECIAL CAUTION. See that the name BECKETT and Trade Mark are on both the Label and Capsule of e ch Bottle.

REDUCED PRICES: Imperial Pints, 1s. 6d., Half Pints, 10d. Tonic Drinks, 1s., 1s. 6d., 1s. 9d., and 2s. 9d. per bottle. Sold by Chemists, Grocers, and Coffee Tavern Companies.

Analytical Report from G. Bostock, Esq., F.C.S., F.A.S., School of Chemistry, Manchester:—"I have made a careful examination of BECKETT'S BEVERAGES. I find them perfectly pure, and free from anything deleterious to health; they are non-intoxicating, and form pleasant and invigorating drinks. The Lime Fruit Syrup, Black Currant, Raspberry, Lemon, Orange, &c., make capital Summer drinks, mixed with either plain or aerated water. The 'Wolseley Liqueur,' 'Winterine,' 'Gingerette,' and Peppermint Cordials are excellent substitutes for Brandy and other spirits, whilst the abundant medical testimony in favour of BECKETT'S TONIC DRINKS—Syrup of Orange and Quinine, Lime and Quinine, and Syrup of Hops—is a sufficient guarantee of their valuable properties."

W. BECKETT, Heywood, Manchester. | London Wholesale and Export Agents: BARCLAY & SONS, 95, Farringdon-street; SANGER & SONS, 489, Oxford-street; NEWBERY & SONS, 1, King Edward-street.

subsequently dedicated the drink to his guardian angel. On the market to this day, it is sold as a dinner aperitif with its medicinal origins forgotten.

TEMPERANCE AND TONICS

The use of alcohol as a tool to tempt the patient to take their medicine can have sinister side effects. From the ruination of eighteenth-century Gin Lane to the era of temperance and teetotalism in the nineteenth and early twentieth centuries, the dangers of alcohol have been much discussed. It was during the Victorian era that the refreshingly bitter, health-giving properties of quinine became valued as an addition to soda syrups and tonic waters aimed at those who abstained.

One of the biggest producers of soft drinks in the 1870s and 1880s was Beckett's Temperance Drinks, which made quinine syrups flavoured with fruits and spices for enhancing carbonated waters. As was common in the period, advertisements included 'official' scientific analysis attesting the drinks' medicinal benefits. Testimonials described Beckett's as '*thoroughly wholesome and pure*', and declared that '*all temperance men to whom Quinine is prescribed should insist upon having [Beckett's] Syrup of Orange and Quinine in preference to all others*'.

It was not just the desirable thirst-quenching properties of quinine that were used to add something extra to temperance tonics; it was used as a deterrent too. In controlled doses, quinine adds a refreshing bitter twist to drinks, but in larger doses it is unpleasant. For

[above] The drunkard's progress, by Nathaniel Currier (1846). Wellcome Collection.

[opposite] From *The Sporting Gazette* (1887). British Library, London, UK.

ECONOMY – Only 2s 6d. per dozen to families.

SODA WATER AND LEMONADE.

HUGHES & COMPY, beg leave to apprise their numerous Friends and Customers, that they are now preparing purest SODA WATER, which contains so great a concentration of the Carbonic Acid Gas, that another atmosphere or degree must inevitably burst to atoms every Bottle employed.

Hughes & Co. have now a NEW SODA ENGINE, on a very improving method; and, as *Practical Chemists*, they have directed their attention for some time to the making of Soda Water on so decided a method, that the Consumer or Physician may depend on Twenty-five Grains of Bi-Carbonate of Soda being contained in every bottle, and no more.

N.B. The making of Soda Water is most simple, as *pure* Water, pure Carbonic Acid Gas, Bi-Carbonate of Soda, great pressure and *precision*, are the only requisites.

HUGHES & CO beg to invite the remembrance of their Customers, that they still continue to sell as extensively as ever their celebrated QUININE PILLS, for Consumption, Weakness, and Debility of the Stomach; and their ROYAL CORN PLASTER (patronised by the Queen) which cures on three applications; nad the FRENCH EMBROCATION, for Gout and Rheumatism.

N.B. Sole Inventors of the QUININE SODA WATER, to which the attention of Physicians and Surgeons in particularly invited.

– Hughes advertisement for quinine soda water, *Bristol Mercury*, 6 June 1835.

those following the temperance code, it was advised to add quinine to wine used for medicinal purposes, to prevent the drinker from enjoying the taste too much and thus being tempted to progress along the slippery slope to drunkenness.

SODA BECOMES TONIC

After Joseph Priestley's 1760s innovation in the manufacture of carbonated water, soda water became a mass marketable product. Within a decade, apothecaries were producing soda waters over the counter, or supplying ready-bottled takeaways, marketed for their medicinal properties.

By the 1820s soda water was being mixed with wine to make refreshing drinks with an emphasis on pleasure rather than medicine. As so often in this story, the timing of the introduction of flavoured carbonated drinks is obscure. Current evidence points to these developing first in the United States, a harbinger of its dominance of the soft drink industry. Between the 1830s and 1850s, syrups of fruit and sarsaparilla were mixed with sparkling water and increasingly sold ready-mixed, in the bottle, marking the establishment of flavoured soda waters.

The first known reference to quinine being added to soda water is in an advertisement in the *Bristol Mercury* newspaper in 1835, for a quinine soda water produced by Hughes & Co., though it seems to have been a short-lived enterprise. The next reference to quinine-flavoured soda water, now termed tonic water, dates to 28th March 1858, in Erasmus Bond's Patent No. 1207, which led to a more successful product. Bond describes '*an improved aerated liquid, known as Quinine Tonic Water*', made from water, carbon dioxide, sulphuric acid and quinine. It was advertised in many local newspapers under the name Pitt's Patent Tonic Water. Very little is known about Bond, but his tonic water

[872]

PITT & Co., 28 *Wharf Road, City Road, London.*—Pitt's patent tonic (aërated quinine) water.

This Aerated Water is the result of extensive chemical research, and has been submitted to several London physicians, from whom it has met with unqualified approval. It is considered by the proprietor to be of sufficient importance to patent, that being the only means by which the public can be protected against fraudulent imitations, and it is now offered under the most flattering testimonials. Its properties are antacid, cooling, and refreshing, combined with all the advantages of Soda Water; it gives strength to the stomach and tone to the whole nervous system, and is especially adapted to persons feeling depressed from mental or bodily excitement, imparting strength to those who suffer from nervous irritation, indigestion, or loss of appetite.

TESTIMONIAL FROM DR. HASSALL.

"Chemical and Microscopical Laboratory,
74 Wimpole Street, Cavendish Square, W.
19th December, 1860.

"I have carefully analyzed PITT'S TONIC WATER. The idea of combining a tonic like quinine with an aerated water is a good one, and the practical difficulties in the way of carrying it out have been entirely overcome in this preparation.

"It is a pleasant, refreshing tonic, and invigorating beverage, strengthening to the digestive organs, and calculated to promote appetite; it is also an excellent restorative to the stomach weakened by any excess or indulgence.

"From its composition and properties, PITT'S TONIC WATER ought to a great extent to supersede the use of soda and other aerated waters."

"ARTHUR HILL HASSALL, M.D., Lond."

Author of the Lancet Sanitary Commission; author of "Food and its Adulterations," "Adulterations Detected," and other works.

The tonic water may be obtained of Messrs. Veillard & Co., Eastern Area of the Exhibition. Numerous medical testimonials may be had on application.

[above] Pitt's aerated tonic water, the first patent tonic water, from *The Illustrated Catalogue of the Industrial Department, International Exhibition* (1862).

was marketed as a medicinal formula rather than as an addition to gin; it was exhibited at the International Exhibition of 1862 with a doctor's testimonial, and advertisements show it was not aimed at fevers and agues but instead promoted as a digestive. It was thus the tonic properties of quinine, not its antimalarial effects, that led to the first tonic water, hence the 'tonic' in its name.

With sulphuric acid as an ingredient, Bond's tonic water may not have been the kindest product on the stomach but the acid had a crucial role in dissolving the quinoline alkaloids, which are less soluble in pure water. By 1870 a new formulation was being produced by Schweppes under the name Indian Tonic Water, with citric acid – kinder to the stomach – replacing Bond's sulphuric acid. The term tonic water was doubtless also a nod to quinine-containing tonic wines.

Tonic water sales quickly reached abroad and were advertised in India and China by 1863. Early references to tonic water as a cocktail ingredient include another 1863 mention, in *The London and China Telegraph*, of its use in Hong Kong for a ginger brandy and tonic water cocktail. In 1867, Samson Barnett's soda and tonic water

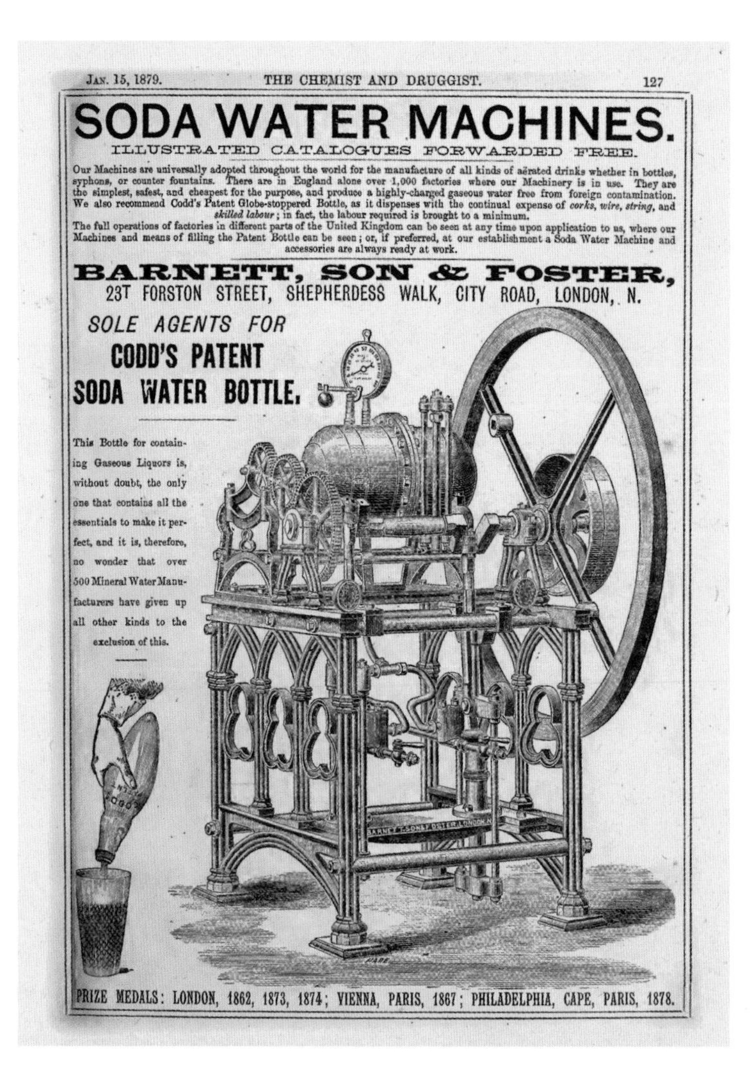
JAN. 15, 1879. THE CHEMIST AND DRUGGIST. 127

SODA WATER MACHINES.

ILLUSTRATED CATALOGUES FORWARDED FREE.

Our Machines are universally adopted throughout the world for the manufacture of all kinds of aërated drinks whether in bottles, syphons, or counter fountains. There are in England alone over 1,000 factories where our Machinery is in use. They are the simplest, safest, and cheapest for the purpose, and produce a highly-charged gaseous water free from foreign contamination. We also recommend Codd's Patent Globe-stoppered Bottle, as it dispenses with the continual expense of *corks, wire, string,* and *skilled labour*; in fact, the labour required is brought to a minimum.

The full operations of factories in different parts of the United Kingdom can be seen at any time upon application to us, where our Machines and means of filling the Patent Bottle can be seen; or, if preferred, at our establishment a Soda Water Machine and accessories are always ready at work.

BARNETT, SON & FOSTER,

23T FORSTON STREET, SHEPHERDESS WALK, CITY ROAD, LONDON, N.

SOLE AGENTS FOR

CODD'S PATENT

SODA WATER BOTTLE,

This Bottle for containing Gaseous Liquors is, without doubt, the only one that contains all the essentials to make it perfect, and it is, therefore, no wonder that over 500 Mineral Water Manufacturers have given up all other kinds to the exclusion of this.

PRIZE MEDALS: LONDON, 1862, 1873, 1874; VIENNA, PARIS, 1867; PHILADELPHIA, CAPE, PARIS, 1878.

From *The Chemist and Druggist* (1879). Wellcome Collection.

"Allow me to offer you any wine you may fancy this hot day, and let me order lunch. There is hock, stern, moselle, claret, champagne; iced soda, seltzer, potass, or tonic water. Ah! you think I am joking – name your wine, and see."

My draught was composed of wine from the Rhine, soda-water from England, and clear ice from the country.

– From *Travels and Adventures of an Officer's Wife in India, China and New Zealand*, by Elizabeth Muter (1864).

machinery is found for sale to regiments in the British Indian Army. A Chinese reference comes from an 1864 book written by Elizabeth Muter, entitled *Travels and Adventures of an Officer's Wife in India, China and New Zealand*. Muter observes that great quantities of soda and tonic water are available (left). Tantalisingly she does not mention if the tonic was ever used for gin cocktails.

In summary, by the eighteenth century, powdered bark, and from the 1820s, purified alkaloids, were widely used in tonic wines and mixed with spirits, sometimes as a general tonic,

St. Raphaël Quinquina.

THE MOST WONDERFUL TONIC WINE IN THE WORLD.

The ST. RAPHAËL QUINQUINA has the largest sale of any similar Wine in France. For Loss of Appetite, Anæmic Conditions, and Nervous Depression, it is the most wonderful Tonic Wine ever submitted. Apart from its marvellous restorative properties, it is also a pleasant and agreeable Wine, and can be used as a beverage with Mineral Water.

Show Cards, Handbills, and all Advertising Matter supplied free.

SOLE AGENTS:

BOWEN & McKECHNIE,
Cross Street, FINSBURY, E.C.

From *The Chemist and Druggist* (1899). Wellcome Collection.

sometimes as a preventative against what we now recognise as malaria. Tonic water became widely marketed in the 1860s and was consumed in the tropics as a healthy and refreshing drink. In 1863 James Henderson noted that in the Shanghai climate '*tonic water is good and wholesome, but care should be taken that it is prepared by a respectable House or Company. Schweppe's tonic water is best, but much is manufactured by speculators in all parts of the world, that will tend to any thing rather than* healthy *tone in the system.*'

However, it is clear that tonic water was consumed for its medicinal 'tonic' effects, for refreshment in hot climes, as a safe alternative to plain water, but not as a preventative against malaria. At this point, before 1868, there is also no evidence it was being mixed with gin.

I want a small Glafs of Gin,
Old Tom Sir?

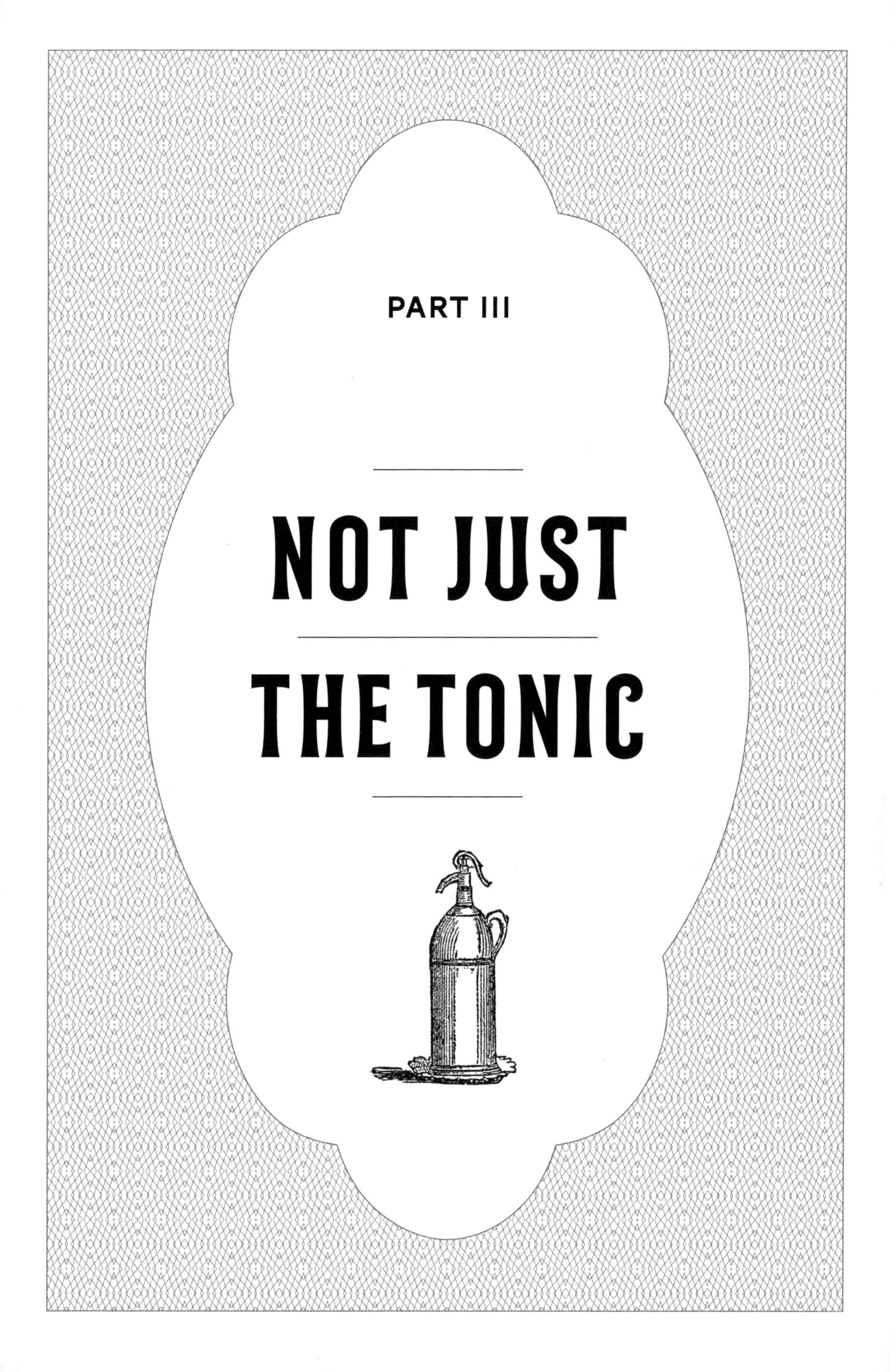

PART III

NOT JUST THE TONIC

6.

ORIGINS OF THE GIN AND TONIC

Modern, and oft-repeated, accounts of the earlier origins of gin and tonic credit its invention to officers in the Indian Army taking their daily bitter quinine dose washed down with gin and soda. Some date this as early as 1825, a mere five years after the first extraction of quinine. Although both quinine and soda water were available in India at this point, it is unlikely that this is the true origin, and no records or references have been found to support this.

A similar story of the origin of quinine-laced alcohol is told in France of Dubonnet, the wine-based aperitif and favourite of the late Queen Mother. This tonic wine was invented in 1846 by Joseph Dubonnet, using a blend of herbs and quinine, and was said to be commissioned by the government to tempt the French Foreign Legion to take their antimalarial medication while abroad.

We believe these origin myths for Dubonnet and gin and tonic are more likely to originate in the long and true history of the admixture of quinine with alcohol as a medicinal tonic, as discussed in Chapter 5.

LEGENDARY ORIGINS?

Quinine as a preventative antimalarial was not introduced to the British Army until the 1850s, when it was taken daily

[opposite] Drinks being served in a colonial setting in India. 1930s illustration.

[previous page] Old Tom was the name of a sweet gin (c. 1830). Wellcome Collection.

as a preventative, not just a cure. The recommended amount was one to two grains per day, in sherry or another alcoholic drink – between 65 and130 milligrams (thousandths of a gram). A modern tonic water contains a maximum of 83 milligrams of quinine in a litre. To achieve any preventative effect from a typical glass of gin and tonic would have required a concentration of quinine five to ten times greater than in a modern tonic. It is not surprising that we have not found any recommendations for the consumption of gin and tonic as a preventative in medical manuals of the day.

Anyone planning to self-dose themselves with modern tonic water should also bear in mind the results of a light-hearted research trial in 2004. This measured quinine blood levels after volunteers

[above] An arrack shop, India (19th century). Wellcome Collection.

[opposite] Quinine was an essential resource for the armies of the American Civil War. 'Before Petersburg – issuing rations of whiskey and quinine' sketch by A.W. Warren, circa 1865. National Library of Medicine.

'Shall I visit Holland by the Damf Schiff [steamer] as the Germans call it? I'll not do that again. It is all dyke and dam, treck schuytt [canal boat] and tobacco smoke. I do not like sipping tea at a lust huis [pleasure garden] to the music of frogs in the green moats around keeping out the miasma with eternal doses of quinine and geneva.'

– From 'Summer Tourists' *The English Journal* (1841)

downed between 500 and1,000 ml (1–2 pt) of tonic water in 15 minutes. Even with this quantity, tests showed only a brief and minimal protective effect against malaria.

QUININE, ALCOHOL AND MALARIA

While quinine has a long history in alcohol as a general tonic, it is from the 1840s that we find the consumption of quinine in spirits specifically as a palatable antimalarial. In the Netherlands, the 1841 *English Journal* complains of a visit to Holland, where gin and quinine must be drunk to keep malaria away.

In the British Army it seems that whichever spirits were most handy were used, most often brandy, whisky, rum, wine or local spirits. An 1863 report on the army in India and Ceylon records that quinine doses were given daily in arrack, a locally distilled spirit.

Rum or 'grog' was the preferred Navy tipple for taking a dose. Fresh water spoiled on long ship journeys, so rum was added to it from the mid-seventeenth century for use by sailors, though with an occasional bottle of gin for officers. In 1740 the rum ration took on the name grog after British Vice Admiral Edward Vernon, who wore a cloak of grogram (waterproofed fabric). Vernon standardised the water-to-rum ratio in twice-daily servings to avoid the risk of drunkenness, part of naval regulations until as recently as 1970. In 1771, naval physicians James Lind (mentioned earlier for his demonstration of the curative effects of citrus for scurvy, see p. 69) and Robert Robertson advised giving Peruvian bark in the rum ration to prevent fever, advice later followed by Admiral Nelson and his fleet doctor, John Snipe, in 1803. By the late eighteenth century there are regular naval recommendations of cinchona bark (then quinine from the 1820s) in grog as a fever cure, and occasional preventative, along with lemon or lime juice for scurvy.

Taking a moderate amount of alcohol was often recommended to help the European acclimatise to the heat and humidity of tropical climates, though there was also a debate about temperance, as reflected in wider society. Dr James Henderson, in his 1863 book *Shanghai Hygiene, or, Hints for the Preservation of Health in China* recommended wine to be safe and refreshing, and spirits '*never taken unless largely diluted*'.

GIN SLINGS AND BITTERS

Gin and bitters were a common combination during the rise in gin drinking, from Georgian England in the eighteenth century to the gin palaces of the early nineteenth century. The bitters part consisted of recipes containing ingredients such as gentian, calamus, angelica, ginger, Bitter Orange and sometimes cinchona bark. Recipes for home-made versions and proprietary blends such as Angostura or Stoughton bitters were available, and these latter were advertised as a tonic for '*all enervating and hot climates*'. A pink gin, gin flavoured with some dashes of Angostura bitters, was the preferred drink of naval officers at sea.

In an account of Panama in 1855 by Robert Tomes, quinine is referred to as a replacement for bitters to create

> [Champagne bitters cocktails] . . . are so supremely good that if he once takes them he will continue to take them and not take the former [water]. I say nothing by way of protest against the frequent practice of drinking quinine cock-tails in which quinine is substituted for bitters and the by no means agreeable but constant habit of freely indulging in quinine pills; for these are excusable, if not necessary on the score of health. It is a melancholy fact that such is the unhealthiness of Aspinwall [Colón] that its inhabitants are obliged to mix medicine with their daily drink, and to pass around their pill-boxes with the frequency of a French snuff taker of the ancient régime. I have been seriously invited, time and again, to drink a quinine cocktail and to help myself out of a proffered box, to a pill or two, which, I need not say, I politely declined.
>
> – *Panama in 1855* by Robert Tomes (1855).

A man grimacing from his bitter medicine (c. 1800). Wellcome Collection.

'hold — I must stop your Grog Jack — it excites thos
and concussions of the Thorax, which accompany Sternutatior
You are in a sort of a kind of a Situation — that You
— Shaved — I shall take from you only 20 oz of Blood — then
and Box of Pills, and
I shall administer
to you a
Clyster.

a Sweat

Jollop

Clyster

By XYZ

Thos. Tegg No 111. — Cheapside — Opposite Bow Chur

JACK, hove down — U

means
t be—
is Draught

Stop my Grog—Belay there Doctor— Shiver my timbers but Your lingo bothers me— You May batter my Hull as long as you like, but I'll be d—'nd if ever You board me with Your Glyster pipe.

Sea Stock

Pig tail

Grog

Rum

True Love Token

P R

ton

a Grog Blossom Fever.

THE MACHINES.

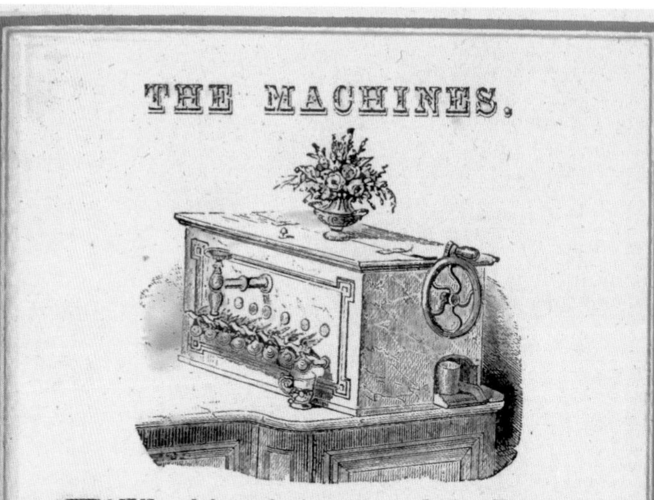

THE celebrated American Soda Fountains I recommend, having on several occasions used them at private residences; for novelty and cleanliness they are unsurpassed. These fountains can be had on hire at a very trifling expense, and the expeditious manner in which they produce the great variety of fruit drinks is a sufficient guarantee for our introducing them to the notice of our readers.

No. 100.—GIN SLING.

USE a soda-water glass. Put 2 slices of lemon and 1 tablespoonful of powdered white sugar or candy, fill up with shaved ice; add 1 glass of gin; shake well, and sip through 2 straws.

D

[left] Soda water machinery and a gin sling recipe, from *The Gentleman's Table Guide* by Edward Ricket (1872). Wellcome Collection.

[opposite] The first known reference to 'gin and tonic' in print. *Oriental Sporting Magazine* (1868). University of Minnesota Libraries.

quinine cocktails for thirsty drinkers with the added benefit of keeping away ague, but it doesn't specify the spirit with which it is mixed. Tomes doesn't think much of them anyway, recommending a champagne cocktail instead (see p. 94). In 1862, Jerry Thomas, an early populariser of cocktails and author of *How to Mix Drinks, or, The Bon-Vivant's Companion* gave a recipe for fever 'drops' consisting of an alcoholic extract of calamus root, zedoary, ginger, dried orange apples and Peruvian bark. His wording '*dose 3 to 4 teaspoonfuls a day*' suggests a medicinal purpose, but its placement in a bartending book with other cocktail bitter recipes merits note. Calisayine cocktail bitters, '*the prince of pick-me-ups*', were briefly advertised around 1879. Made with cinchona for adding to gin cocktails and soda water, they do not appear to have become the basis of a popular drink.

We thus have plentiful evidence of gin drinks with a bitter profile, sometimes based on quinine, but with no evidence of dilution by water, sparkling or still.

there was plenty of betting, and our modest fiver went on *Polly*, more for the sake of backing her rider than thinking of what class she was. Loud cries of "gin and tonic," "brandy and soda," "cheroots," &c., told us the party was breaking up for the night, and we wended our way home (only a short distance from the mess, luckily), feeling certain we could lay 2 to 1 we named the winner of each race on the morrow, only that it would be a very rash bet to make.

In contrast, we also have evidence of a drink containing gin and sparkling water, but lacking quinine. This is the gin sling, recorded in bar books, articles and even a few poems from at least 1829. The sling contains a refreshing mix of gin, soda (or water), sugar, ice and a slice of citrus; everything in a gin and tonic but the quinine. The taste for both gin and bitters and gin slings may have naturally merged into the classic cocktail we now know, but this next step is frustratingly poorly documented.

GIN AND TONIC AS A COCKTAIL

The first known reference to gin and tonic as a bar cocktail is in the Anglo-Indian *Oriental Sporting Magazine* in 1868, a decade after the first patented quinine tonic water. The term was evidently a familiar phrase in India, being called out by attendees of a horse race at Sealkote (Sialkot) as they finish for the evening (below right):

Note here the double contexts of the army and horse racing; both are important in the early history of gin and

'Careful officers have a cup of tea about five in the morning, then, perhaps, about nine or ten, oatmeal porridge, fried mullet, strawberries or sliced tomatoes – perhaps a light lunch of cold chicken, perhaps none; perhaps sherry and bitters at the club – the comfortable Wheler Club; perhaps a gin tonic well iced – anything to sustain Nature until eight o clock dinner, when the cautious drink claret or a little sherry, the economical subaltern his dearly beloved beer; later in the evening a cup of coffee, the best of all comforters – the cigar, then a game of billiards, a hand at whist or a cosy, comfortable armchair, a soft pleasant light to read the newspapers, periodicals or a novel by Trollope, Whyte, Melville or some sensational dashing lady writer – then to bed. This ascetic life is not led by all.'

– Indian Medical Notes, *The Medical Press and Circular* (1875).

'This last ride had made us considerably thirsty, so we made our way to a big tree at the edge of the tank, where the liquor-boxes were. Most of us had dismounted but the old Judge preferred imbibing on horseback; he got a glass of gin-tonic, and the twinkle in his eyes showed that he was going to enjoy it when, all of a sudden, there were shouts of a made buffalo charging. I believe it was Gibson who gave the first alarm. The big Judge on the big horse, was the first to draw the buffalo's attention, and as he charged Bainbridge did not wait to look for his glasses or finish his peg, but dropping his gin-tonic he soon cleared out. The buffalo, a very gaunt-looking specimen, was perfectly mad, for, charging and clearing us all round, he next made for the elephants.'

– Reminiscences of *Twenty Years' Pigsticking in Bengal* by 'Raoul' (1893).

tonic. Other references in this early period show that it was enjoyed throughout the 1870s and 1880s as a pleasurable, rather than medicinal, drink to relieve the heat of tropical climes, and was particularly associated with the English. A South African newspaper, *The Lantern*, in 1881 poked fun at an English man recently '*imported*' who is recognised by his swagger and stare, while sipping on a gin and tonic. Though '*he thinks himself a swell-ah*', he is becoming a down-and-out because he cannot get a job, and '*prospects*' seem quite '*dre-ah*'. Some references are ambiguous: does the term a '*gin tonic well iced*', in *Indian Medical Notes* of 1875, refer to the cocktail, or simply to gin *as* a tonic? The ice suggests the cocktail.

In an 1882 news article from the Aberdeen Journal describing 'A morning with bobbery pack' (a mixed pack of hunting dogs) the anonymous author describes an episode starting with '*we were drinking gin and tonic under the old tree on the Calcutta maidan*'.

Hugh Wilkinson enjoyed '*tin-gonics*', in other words gin and tonic, in his 1883 book *Sunny Lands and High Seas: A Voyage in the S.S. Ceylon*, and an 1893 biography entitled *Reminiscences of Twenty Years Pigsticking in Bengal* refers to the memory of Judge Bainbridge, who in 1879 '*got a glass of gin-tonic, and the*

> We were drinking gin and tonic under the old tree on the Calcuttta maidan after a hard fought game at polo, when a friend from government house said: 'I say L., you are having great sport with your bobbery pack, we hear, will you give us a morning?' 'Yes' I said 'first rate fun, but it is not hunting mind, only we generally kill something'.
>
> – A morning with bobbery pack, *Aberdeen Journal* (1882).

WHY DOES GIN AND TONIC WORK?

Matthew Hartings of American University, Washington DC, describes the chemistry behind why gin and tonic are a match made in heaven. Hartings says flavours work best when the molecules that tingle our taste buds have a complementary shape. Quinine, an alkaloid molecule, and the molecules of juniper, which are used to flavour gin, both have similar ring structures and are attracted to each other, aligning and combining to create a taste that is greater than the sum of its parts. This combination explains why a new flavour profile is created, rather than a simple juniper–quinine taste.

[opposite] Hunting a panther in India (c. 1840). Wellcome Collection.

twinkle in his eyes showed that he was going to enjoy it'.

Consumption of gin and tonic did not guarantee good health. In 1875 one unfortunate sailor is recorded as dying of cholera in Bombay, spending his shore leave enjoying a gin and tonic before a final and fatal supper. The 1875 novel *A British Subaltern* depicts a feverish patient demanding a gin and tonic:

'*"I don't think I am long for this world", observed a very unhealthy-looking ensign, who had been laid up with an attack of fever for the last few days. "A gin-and-tonic, and look sharp with it!" shouted this gentleman on the appearance of one of the Kitmutgars, or black servants, at the door of the ante-room.*

"No; I think you had better make your peace with your Maker. I don't think that you are long destined to remain in this 'vale of tears'", was the consoling rejoiner of another equally unhealthy-looking subaltern, emptying as he did so a large tumbler of iced liquor.'

LEMON AND LIME

A slice of lemon is a crucial ingredient in traditional gin and tonic. Lemons have been available in Britain since Elizabethan times. By the time gin and tonic evolved, over twenty million lemons were sold in London each year. Getting them from the Mediterranean to Britain was a complex business. *Household Words* in 1854 described the requirements:

'*The beautiful ships, of the noble steam vessels engaged in transporting them from foreign lands to these shores: of the railway-trains employed at certain seasons, to whisk the cooling cargoes from Southampton to London, while their consumers are sleeping in their beds: of the large piles of massive warehouses required to store, to sample, and to sell them by auction: of the mean squalor and desolation of the great retail orange-mart in Duke's Place: of the thousands of men, women, and children who draw a subsistence from their sale in the streets, in steamboats, at fairs, in theatres, or wherever people congregate.*'

WHAT WE KNOW ... AND DO NOT KNOW

The twenty years following Bond's tonic water patent of 1858 is the crucial period in the invention of gin and tonic. Tonic waters were being marketed in Britain by the early 1860s, with both the bottled drinks and the technology to make them exported to India. Perhaps equally importantly, by the 1880s cheap and dependable supplies of quinine – a key ingredient of tonic water – became available from the India, Sri Lanka and Java plantations. All the nineteenth-century references to gin and tonic, from 1868 onwards, are from India and many but not all have military connections. By 1870, Schweppes used the appropriate name Indian Tonic Water for its tonic

Our Last Importation
Let me introduce you to a fell-ah,
La de dah, La de dah,
Who thinks himself a swell-ah,
La de dah, &c.
He's just come by the steam-ah,
La de dah, &c.
And is a perfect sweam-ah,
La de dah, &c.
He puts up at the Masonic,
La de dah, &c.
And sips his Gin and Tonic,
La de dah, &c.
His outfit is quite new-ah,
La de dah, &c.
'Tis astonishing to view-ah,
La de dah, &c.
He toddles down the street-ah,
La de dah, &c.
And ogles all he meets-ah,
La de dah, &c.
But in Cape-Town 'tis a bo-ah,
La de dah, &c.
It's cads he could not know-ah,
La de dah, &c.
What to do he is not shu-ah,
La de dah, &c.
A clerk he can't endu-ah,
La de dah, &c.
He came out for 'a Commission,'
La de dah, &c.
But he finds there's no admission,
La de dah, &c.
So he's waiting a remittance,
La de dah, &c.
Till his landlord gives him quittance,
La de dah, &c.
Till his clothes are getting seedy.
La de dah, &c.
And he's altogether weedy,
La de dah, &c.
And our noble swell I fea-ah
La de dah, &c.
Has prospects somewhat dre-ah

– A poem about a gin-swigging British immigrant, *The Lantern*, South Africa, 1881

water. Yet there is nothing to suggest a medical purpose in its consumption, rather it is the refreshing properties of gin and tonic in the tropics that come to the fore.

The medicinal properties of quinine are, however, relevant to the origins of gin and tonic. The long history of quinine as a tonic for general health surely inspired Bond to develop his tonic water and encouraged a ready market for the product. Equally, the long history of quinine in alcoholic drinks, whether in tonic wine or in spirits, must have suggested to the first consumers of gin and tonic that the combination was a plausible one. And, even if not consumed specifically as an antimalarial, the healthy connotations of quinine doubtless contributed to its reputation as a suitable drink for hot climes.

As more sources come online and can be easily searched, the history of gin and tonic will undoubtedly be refined. The archives of manufacturers such as Schweppes are also an under-studied resource. These may reveal more about the timing and export of sales of tonic water to India, undoubtedly the key place for the origin of gin and tonic. Letters, account books, bar bills and local newspapers may also reveal more about the elusive creation and spread of gin and tonic.

Botanical Illustration of a lemon from *Köhler's Medizinal-Pflanzen* (1887).

Citrus Limonum Risso.

ICE

Ice is an essential ingredient in many beverages, including gin and tonic. In today's era of domestic refrigerators, it is hard to remember that ice was an expensive commodity until the mid-twentieth century. Before modern technology, ice was harvested in far-distant mountains, or even from abroad. A complex 'cold chain' was required to transport it to where it was needed, with insulated transport and storage to keep it below freezing from the time and place of harvesting to the time and place of consumption – perhaps six months and several thousand miles distant.

Ice has been harvested and stored for many thousands of years. A cuneiform tablet records the building of an ice house for storing ice, at Mari in Mesopotamia some 4,000 years ago, lined with tamarisk branches for insulation. It would have been filled with ice from the Zagros mountains that flank Iraq to the north and east. Records from China 3,000 years ago, and from classical Greece and Rome, also refer to ice and snow being sold in cities. Snow or ice could be mixed with drinks or packed around a jar to cool the liquid. In Europe ice houses appeared in the medieval period, in monasteries and castles, and became a standard accoutrement for the country houses of the landed gentry in the eighteenth century. They were often built next to a lake, reflecting a colder era in which large amounts of ice could be harvested even in southern England. Like many others, the eighteenth-century ice house at Kew Gardens survives today. Seventeenth-century travellers to Persia and Turkey saw elaborate constructions for the preservation of snow and ice, and encountered sherberts, perhaps the ancestor of the European tradition of sorbet.

Two early nineteenth-century innovations transformed the availability of ice for cooling, placing in drinks and storing food. Throughout the century, large-scale exports of New England lake ice left by boat for port cities as far away as India, the Caribbean and the British Isles, and to China and Japan as these opened to Western traders. Much of the trade was masterminded by Frederick Tudor (1783–1864), a Boston trader known as the Ice King. To keep the cold chain intact, large ice houses, sometimes funded by grateful citizens, sometimes by Tudor, were built, as in Calcutta (today's Kolkata, India). The scale of the trade was huge, with domestic consumption in the United States 5 million tons by 1880, and exports to India reaching a peak of 146,000 tons in 1856. In India the railways opened up ice supply to inland areas, and thence to domestic ice boxes. Using these the *abdar*, the servant in charge of drinks, would prepare cool drinks for the household, both for pleasure and as an essential medical aid for those ill in a tropical climate.

The earliest ice-making machines date to the 1830s and initially used dangerously corrosive acids. By the 1860s vapour refrigeration machines, the ancestor of today's refrigerators, were in use although it was several decades before the sea-borne trade in ice ended. In Europe and North America by the beginning of the twentieth century, ice had changed from being a luxury to a necessity. An icebox, used for storing food and regularly topped up with fresh ice, was an emblem of the United States; in Europe, cellars and pantries were instead favoured for food storage. Refrigerators became widely available to industrial and large commercial premises by the end of the nineteenth century, but only entered homes from the 1920s, becoming widespread in the United States after the great depression of the 1930s and from the 1960s in Europe. Machine ice was not only more reliable in its supply but also, for use in drinks, less likely than river or lake water to be contaminated by disease-causing organisms and industrial by-products. The almost universal use of ice in cocktails and soft drinks owes everything to its easy availability, as well as its property as a low-cost filler to extend a drink.

Refrigeration using Ether: Siebe & Harrison's ice-making machine. Capable of making 269 gallons of spring or river water into blocks of solid ice. Shown at International Exhibition, London, 1862. Copperplate engraving. Wellcome Collection.

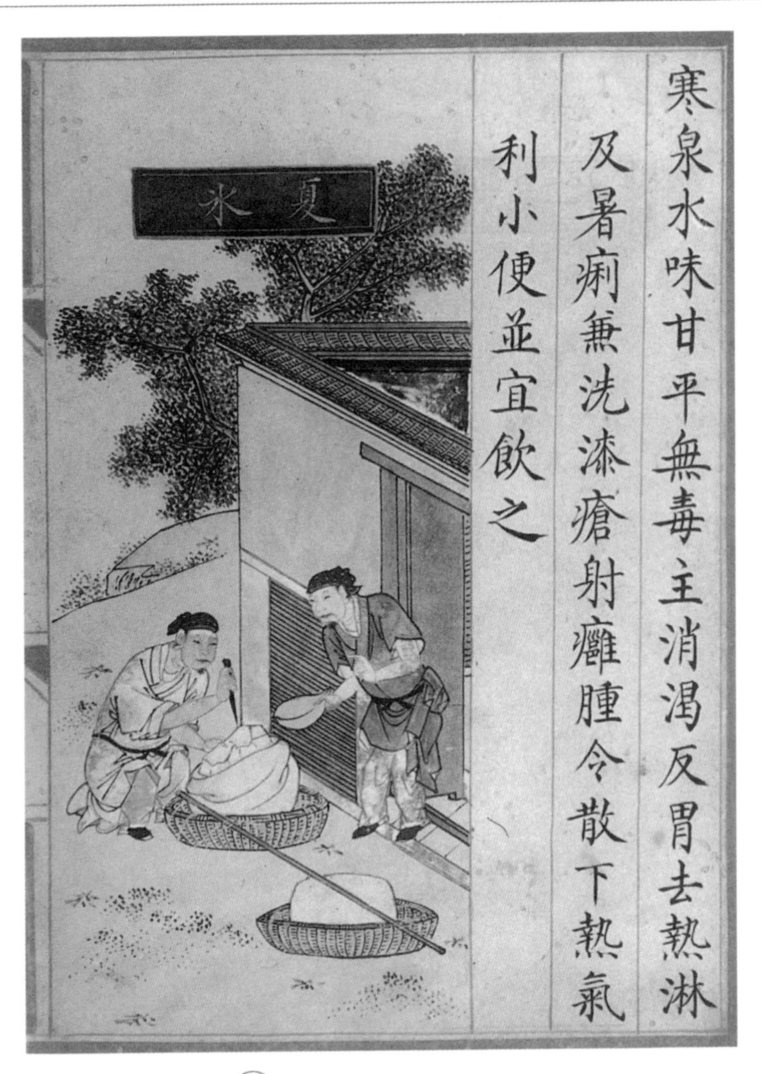

Summer ice from the *Shiwu Bencao* (*Materia Dietetica*), a dietetic herbal in four volumes dating from the Ming period (1368–1644). Wellcome Collection.

Harvesting ice from the Great Lakes of North America, c. 1903. Library of Congress.

George Cruikshank fec

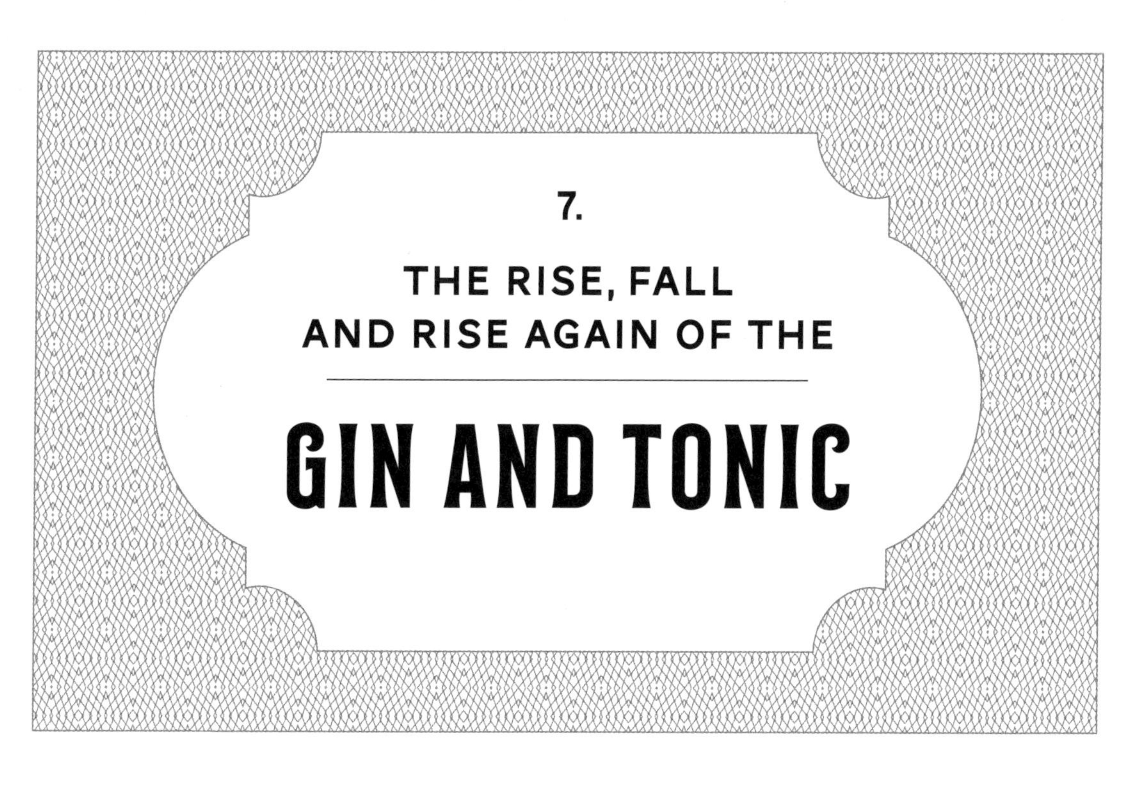

7. THE RISE, FALL AND RISE AGAIN OF THE GIN AND TONIC

We have seen how tonic water appeared in the 1860s as a healthy 'tonic', and how the uniquely refreshing properties of gin and tonic were discovered in the tropical climes of India by 1868. Since then, both components of gin and tonic have had a chequered history, recognised as a distinctly English long drink by the 1920s, spreading to the United States in the 1950s but going into the decline in the 1970s as tastes in alcohol changed. The last decade has seen a remarkable revival in artisan gin – rapidly followed by a similar boost to tonic water.

A SHORT HISTORY OF GIN

The British have not always drunk gin. Its origins lie in the medieval Low Countries, where genever originated as 'jenever', named after the juniper berry, which is the defining ingredient of gin. Starting as a flavoured brandy, distilled from wine, in the sixteenth century it became the grain-based drink that is still distilled in the Netherlands today.

British soldiers met genever in the Thirty Years War (1618–1648), drinking it for Dutch courage, but as an expensive import it did not become widely popular. However, in the seventeenth century it inspired domestic production of aqua vitae, similarly distilled from grain and juniper berries, and which became known as gin, anglicised from genever.

[opposite] The bar in a gin palace, by George Cruikshank (c. 1842). Wellcome Collection.

As a locally made product, prices fell and production – and consumption – increased, although juniper was replaced by oil of turpentine (from pine or larch trees) in cheaper versions. By the eighteenth century beer production was falling and gin production had greatly increased.

England came late to spirits compared to the European mainland or Scotland, but soon made up for it. By the 1730s eleven million gallons of legal gin (approximately 50 million litres / 13 million US gallons) were produced each year in London alone, equivalent to fourteen gallons (approximately 63 litres / 17 US gallons) per person. Something of a gin panic ensued, with gin held responsible for

[above] The gin seller and the three beggars. Etching by Jan van Ossenbeeck, 17th century. Rijksmuseum.

[opposite] *Gin Lane* by William Hogarth (1751). Wellcome Collection.

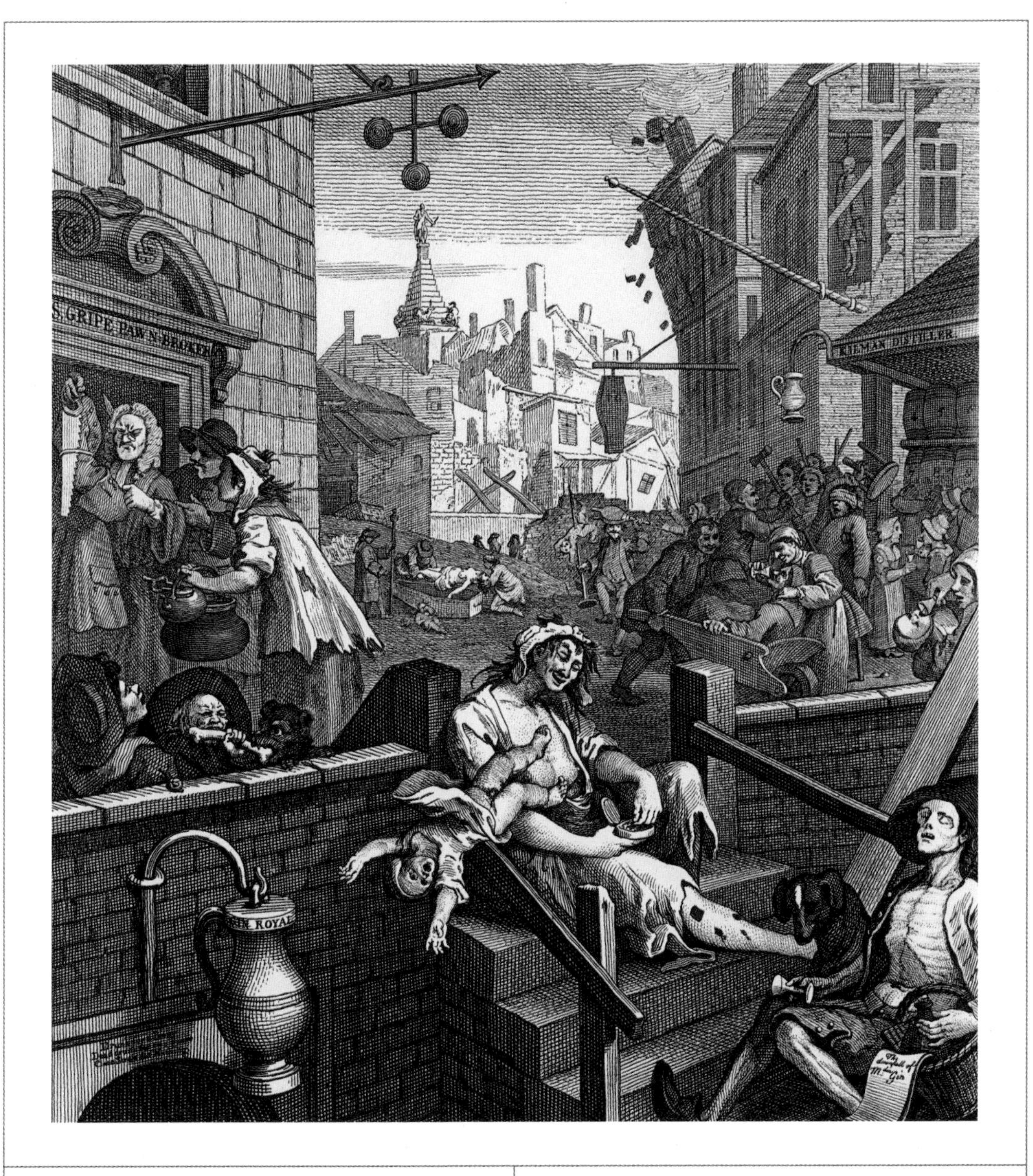

dissolute behaviour among the working classes, vividly illustrated by William Hogarth in Gin Lane. A series of laws designed to regulate the gin trade did little except drive it underground, leading some historians to compare it to more recent attempts to regulate drugs in modern times.

By the mid-nineteenth century the gin palace, a form of pub, had become widely established in England's cities. Advances in building materials enabled '*a splendid edifice, the front ornamental with pilasters, supporting a handsome cornice and entablature and balustrades, and the whole elevation remarkably striking and handsome... the doors and windows glazed with very large squares of plate glass, and the gas fittings of the most costly description*'. Typically featuring large open spaces and a long, polished counter, a gin palace lacked seating and could serve thousands of customers in a day.

In the 1830s continuous still techniques were developed that allow the production of alcohol of consistent quality. It is not a coincidence that some of today's major distilleries, such as Gordon's and Beefeater, were founded at about this time. They produced a drier gin than before, typically flavoured with ten to twenty botanicals and recognisably the modern dry London gin that we mostly drink today.

THE EVOLUTION OF THE COCKTAIL

Gin caught on more slowly in the United States. Newspaper advertisements from the 1870s promote both Dutch and London gins for their medicinal properties although, as during the Prohibition period, one suspects they may also have been enjoyed for their pleasurable qualities.

America had a long tradition of cooling drinks, particularly in the hot and humid American South, home to the julep. There is a fine line between these and the true cocktail, which strictly defined is as a mixture of one or more alcoholic drinks, sugar and bitters. Both long drinks and cocktails were served in the American Bar, a distinctive contribution of the newly urbanising United States to the history of spirits.

Bartenders such as 'the Professor' Jerry Thomas, of the Metropolitan Hotel, New York, turned mixology into an art form. Thomas opened the first American Bar in Europe in London in 1868, and an American Bar is today still a feature of smart hotels worldwide. Thomas also

[above] The Gin Juggarnath, from *My Sketch Book* by George Cruikshank (1834). British Library.

[opposite] The 'Blue Blazer', a fiery combination of whisky and boiling water that demonstrated the bartender's art. From *How to Mix Drinks: Or, The Bon-vivant's Companion* by Jerry Thomas (1862).

wrote one of the best-known handbooks on the cocktail, still an inspiration to today's mixologists. In the 1887 edition of his *The Bar-Tender's Guide, or, How to Mix All Kinds of Plain and Fancy Drinks*, most of his gin recipes are based on Dutch genever gin (Holland gin).

The story of gin in the United States is of course interrupted by the Prohibition era of 1920–33, when drinking joints known as speakeasies thrived, selling illegally made 'bathtub gin'. Throughout this period, pharmacies sold alcohol for medicinal purposes, blurring the line between drinks for pleasure and medicine. As late as 1933, Jack Grohusko's *Jack's Manual* highlights the medicinal properties of Holland gin (genever), for its '*purifying effects on the kidneys if taken in moderation*'.

A TROPICAL DRINK

As we saw in Chapter 5, from the 1880s to the 1920s gin and tonic was something of a shorthand for the tropics. The rare references to consumption in Britain invariably have a tropical origin. In 1881, a disgruntled letter writer to *The Sporting Times* laments the lack of tonic water in England, compared to India.

In fact, as we have seen in Chapter 4, tonic or quinine water had been widely advertised since the 1860s, but the notion of mixing it with gin had evidently not caught on in Britain by the 1880s.

To the editor of the 'Sporting Times'
Dear Sir, – Why, in the name of all that is mysterious, is not tonic water procurable in England – the land that boasts almost every kind of mineral water? I have passed a good portion of my life in the land of India and have now just returned after a six years' sojourn in that country. India has its blessings, as well as its – well the reverse of blessings – but one of its greatest enjoyments is the pleasant 'peg'. We have out brandies and sodas, as well as other drinks procurable in this delightful country, but we fail to find here the drink most patronised in India, viz., gin and tonic. Of course, the gin is to be had, but where the tonic? This question I want to ask you, you may be able to answer it or if not, by kindly inserting it in the Pink 'Un it may catch the eye of an old Indian, and he may be able to inform me through your paper if I can obtain, and if so, where, the drink required. I am, sir, yours faithfully, ANGLO-INDIAN. April 28th 1881.

– A letter to the editor of *The Sporting Times*, London (14 May 1881).

In June 1914 – hindsight lends pathos to the date – *The Times* published a whimsical review of the geography of drink that shows familiarity with many forms of mixed drink, whether consumed in Britain or overseas. *The Times* had firm views on drink and climate: '*many Britons, it is to be feared, find their graves many years before they need because they insist on carrying their British habits to the tropics, but roughly it may be said that shandygaff [beer and ginger beer mixed] ceases to be tolerable within the gin-and-tonic degrees of latitude*'.

Billie's Bar, 56th Street and First Avenue, Manhattan. Photograph by Berenice Abbott (1936). New York Public Library.

GIN AND TONIC REACHES EUROPE

The gin palace was not the natural home for cocktails such as gin and tonic. However, it was not far away, in the clubs and hotels favoured by the British establishment, that long drinks and cocktails were becoming popular in the mid-nineteenth century. They gradually replaced the cups or lower-alcohol punches previously drunk at a middle-class social gathering. While we might think of the 1920s and 1930s as the cocktail age, gin historian Richard Barnett points out that the key ingredients – bitters, vermouth and soda – had been developed in the eighteenth century and came together to form the cocktail in the nineteenth. The great chef Alexis Soyer was a fan of the cocktail, and his Universal Symposium of All Nations, built to feed the throngs at the Great Exhibition of 1851, featured The Washington Refreshment Room with forty cocktails on offer. However, *Cooling Cups and Dainty Drinks*, published in London in 1869, still devotes about a third of recipes to cups and punches, and just a few pages to cocktails including '*that nectarous compound "Mint Julep"*', the Locomotive, and the Gin Cocktail.

The progress of gin and tonic to the British Isles from its Indian roots is even more mysterious than its tropical origin. Richard Barnett notes that concrete evidence for the transfer of gin and tonic is scarce, but suggests it reached the cocktail menus of London's better hotels by the early twentieth century. Our own research, based on the less exclusive pages of the daily newspapers, points to the mid-1920s as the first time that gin and tonic is referred to as anything other than a tropical drink in the British press.

It is striking how often it is picked out as the drink of choice for hot weather in Britain, perhaps simply reflecting its truly refreshing nature. In 1925 *The Bystander* features Cyril, a '*man about town*', inviting his friend to escape the heat with '*an iced gin and tonic consumed slowly in my club hard by*'. Cyril is also a betting man, and perhaps the strongest association of gin and tonic in the press at this time is with horse racing. It might not be a coincidence that a highly successful racehorse in the 1920s was named Gin and Tonic. Sadly, throughout the 1920s gin and tonic is often cited in court cases, usually featuring drunken drivers who have had anything from one to four drinks (sometimes doubles) before setting off, and who have been surprised at the effects on their driving skills.

It would be tempting to connect the twentieth-century arrival of gin and tonic to England with the return of retired army officers from India, but this would not explain why it took so long for a drink known in India in the 1860s to become established in the home country. It is also likely that contemporary observers of the social scene would have commented on the connection. Perhaps the gin and tonic wasn't thought suitable for the British climate? The social history of the bar in this crucial period also deserves closer scrutiny from historians.

In the 1930s the first of many associations of gin and tonic with golfing appear, with the *Green 'Un* (a weekly sporting newspaper) recommending '*a small gin and tonic before playing*'. Sales of tonic water were doubtless encouraged by Schweppes' first advertising campaign specifically for gin and tonic, in 1933. As we will see in its later promotion of tonic water in the United States, Schweppes did not stint on its marketing, and this may well have had a significant effect on raising awareness of gin and tonic as an everyday drink. Intriguingly, the earliest recipe for gin and tonic that we have found in a bartenders' manual is in Frank Meier's *The Artistry of Mixing Drinks* (Paris, 1936), specifying a '*split of Schweppes Indian Tonic Water*' as the mixer. Did Schweppes' promotional efforts extend to continental Europe?

WARTIME SHORTAGE

The rise of gin and tonic was interrupted by the Second World War, when the Japanese invasion of Indonesia in 1942 cut off the supply of quinine from the cinchona plantations in Java. Although quinine continued to be produced in India and Africa, supplies were limited. The soft drink industry in Britain was put under government control to ensure wisest use of resources. Under the aegis of the Soft Drinks Industry (War Time) Association, some factories were closed in 1943 and production shifted emphasis from table waters such as tonic to concentrated fruit cordials that could be transported more efficiently.

In North America the war had similar effects. Canada Dry had launched its first Indian Tonic Water in 1938, targeting a Canadian bar trade that relied almost entirely on imported tonic water. Production ceased in 1941 as quinine was channelled to medical uses, restarting in 1945. In the United States the medical authorities resorted to asking pharmacists to check their store cupboards for unused quinine powders. Huge quantities were recovered from pharmacists in the northeast United States alone. This is not usually a malarial zone, and the quantity of quinine stocked there is another reminder that the tonic effects of quinine were considered – even in the 1940s – to be beneficial for health well beyond malaria. The United States also undertook the systematic exploration and exploitation of the wild cinchona forests of South America, re-initiating the old trade that was so important before the Java plantations came on stream.

Piccadilly Circus, London, photographed by an airman of the US Army Air Force during the Second World War. The lights stayed off until 1949. Imperial War Museum (FRE 10686).

Quarterly
SUPREME
GIN
BILE BEANS
KEEP YOU HEALTHY
HAPPY AND SLIM
WRIGLEY'S
WRIGLEYS
P.K
WRIGLEYS
SPEARMINT
FOR VIM AND VIGOUR
HAIRDRESSERS
HAIRCUTTING
MANICURE
GREYS
GX 757

POST-WAR EXPANSION

When the lights went back on in Piccadilly Circus in 1949, long after VE Day, they brought back to life the long-standing illuminated signage for Schweppes Tonic Water, part of the famous night-time lights. As the leading producer of tonic water, Schweppes was to play a major role in the post-war spread of gin and tonic. The wartime soft drinks industry scheme finally came to an end in 1948, sugar controls ended in 1953 and a period of expansion started for soft drink manufacturers.

Before and during the Second World War gin and tonic was considered across the Atlantic as a distinctively English drink. In 1930 Rudyard Kipling, a great fan of gin and tonic, wrote to his son-in-law from Bermuda: '*these folk ... are so benighted that they do not know "gin, tonic and bitters" as a drink. I have asked for it. The bar-man now dispenses it as a "Kipling"*'.

In 1944 the American journalist Noel Busch noted '*English people eat oily little fish, lukewarm mush, and other such stuff for breakfast. This suits them perfectly. They also consume strange drinks, such as gin and tonic, which serves to keep the swamp fever out of their bones.*' Gin and tonic was largely a post-war introduction to the United States, coming on the back of increased promotion of tonic water.

Canada Dry's Indian Tonic Water was renamed Quinac (rhyming with Whynac) in 1949, in part due to a United States government edict that quinine water was not a true tonic in the medical sense. In 1952 Quiniac sold 935,000 cases in the United States, about 80 per cent of the tonic water market. Gin companies found their sales increased, as marketing of tonic increased the demand for gin and tonic, and the drink started to appeal outside the solidly gin and tonic centres of New York and Washington. Bartenders were a particular target for marketing campaigns, as gin and tonic was easier to prepare than more complex beverages.

It is not a coincidence that Schweppes launched its own tonic water in the United States at the same time. In 1953 it entered into a franchise agreement with Pepsi-Cola. By 1955 some seventy million drinks were mixed with Schweppes Tonic in the United States, bring a welcome supply of US dollars to Britain. Commander Edward Whitehead was sent to the United States to head the marketing effort, and his nattily attired figure and splendid facial hair were heavily featured in Schweppes advertising of the fifties and sixties on both sides of the Atlantic, doing much to promote gin and tonic as a sophisticated drink.

Tonic water was often found to be too bitter for a straight drink, and in 1957 Schweppes introduced Bitter Lemon as a soft drink. Its inspiration lay in drinks such as Orangina, made with citrus fruit pulp. Finely ground lemon fruits give Bitter Lemon both its aroma, derived from the essential oils in the peel, and its cloudy appearance. Television advertising helped it to sell more than thirty-six million bottles in its first eight months, outstripping a Bitter Orange drink launched at the same time. Advertising promoted the sweeter Bitter Lemon to women. Although something of a niche product today, Bitter Lemon – and plain tonic water – is a popular standalone drink in many warmer climates, and similar products to Bitter Lemon are made by several manufacturers.

DECLINE

The big story in soft drinks in the twentieth century was, of course, the rise of colas. Cola drinks were initially developed in the United States, most famously in the form of Coca-Cola. Invented in 1886 by Atlanta pharmacist John Pemberton, Coca-Cola was promoted as much for its health-giving properties as for its refreshing qualities. Pepsi-Cola (1893) and Dr Pepper (1885) were part of the same wave of carbonated

drinks that served the thriving soda fountains of the time, before diversifying into bottled form and going on to mould American popular culture. Colas made little inroad into Europe until the 1950s, when manufacturing and marketing followed major wartime efforts to supply soft drinks to soldiers in the field. The dominance of colas and other American soft drinks, especially once low-calorie forms were introduced in the 1970s, shifted consumption and marketing away from the traditional table waters such as tonic and lemonade, which was until the 1950s by far the most popular carbonated drink in Britain.

In the 1970s tonic's partner, gin, suffered from its middle-class associations, with gin and tonic becoming the cartoonist's shorthand for the golf-club bore. Furthermore, the limited suite of dry gins on the market did not approach the diversity of taste and price so easily available for whiskies. gin and tonic continued to be enjoyed by many, but did not catch the imagination of the serious hobbyist or of the younger crowd. By the mid-eighties gin cocktails such as the martini (gin and vermouth) had also lost their appeal.

[above] Soda fountain in Staten Island, New York, c. 1920. New York Public Library.

REVIVAL

The marketing of tonic water in the 1950s United States did much to establish its domestic gin distilling industry; in contrast, it has been the development of artisanal gins in Britain and North America that has revived the tonic water industry in the last two decades. Bombay Sapphire was the first of the new premium gins, leading the way in 1988, with a striking bottle and an emphasis on quality botanical ingredients. Major producers such as Gordons and Plymouth Gin have followed suit but the greatest revolution has been the growth in small-scale production in copper pot stills.

Pioneered in New York State in the early 2000s, artisanal production began in Britain in 2009 at the Sipsmith distillery, located in west London not far from Kew Gardens. The founders had a two-year battle with Her Majesty's Revenue and Customs to receive a licence to distil: their concept of small-scale production fell foul of eighteenth-century laws aimed at banning the production of moonshine. Once the legal barriers were lowered, artisanal distilling took off, with hundreds of micro-distilleries making gin in the United States and Britain today.

The soft drink industry took a while to respond to the revival in spirits – which benefited vodka too – but in 2005 a new company, Fever-Tree, launched its first product, Indian Tonic Water. Originally available only in the more upmarket supermarkets, by 2008 it was available nationally. Fever-Tree has continued to flourish, now offering seven variants on tonic, including lemon, orange, elderflower and cucumber. All contain quinine, sourced from plantations in the Democratic Republic of Congo. Other manufacturers such as Fentimans, Double Dutch and East Imperial have followed suit with premium tonics, and in 2017 Schweppes relaunched its soft drinks in pear-shaped bottles in a nod to its beginnings more than 200 years before. In 2018 sales of tonic water in the United Kingdom totalled an astonishing £687 million, not far off half the total sales of gin for the year, and about a third greater than tonic water sales in 2017.

For all its success, tonic water has only recently seen off a threat to its very name. The question arose in the European

Kew's own gin and tonic.

Union as to whether the term 'tonic' implied health benefits that did not exist. After representations from manufacturers in Britain and elsewhere, the 2019 *Official Journal of the European Union* recorded agreement that tonic is, in this case, a generic term for 'Non-alcoholic carbonated beverage containing the bittering agent quinine' and can continue to be used.

THE FUTURE?

These are complicated times for the drinks and soda industries, with greater public awareness of the effects of alcohol and sugar on health. In a nod to its origins as a medicinal drink, tonic water may be well-placed to ride out these changes.

The gin and tonic is now a major feature of some bars and pubs, served in the balloon-shaped Copa de Balon glass first popularised in the Basque country. The emphasis is on visual presentation and customisation through botanical garnishes, rather than on alcohol consumption.

In 2017, the Office for National Statistics produced its regular report *Adult Drinking Habits in Great Britain*, revealing that alcohol consumption was at the lowest level on record, highlighting that sixteen to twenty-four year olds were the age-group least likely to drink. The reasons behind this shift are complex, but market trends suggest that younger generations are turning away from alcohol and seeking alternatives. Some experts have suggested the curation of online personae in the Instagram-age is promoting a new wave of temperance.

This trend is linked to the rise of non-alcoholic alternatives for interesting adult soft drinks exemplified by Seedlip, who produce flavoursome 'spirits' distilled from vegetables and spices, alongside the spectrum of artisanal tonics, bitters, syrups and sodas now available on supermarket shelves. It seems that tonic water is going from strength to strength, embraced from both sides of the market: the gin lovers *and* the teetotal mocktail aficionados out to have a good, but healthy, time.

The last chapter in the story of tonic has not been written!

KEW GIN AND TONIC

Aromatic juniper berries (*Juniperus communis*) are the defining flavouring of gin, but other botanicals are added to create the unique flavours found in different brands. Kew's botanically inspired gins were first launched in 2017, crafted by the London Distillery Company using 42 different botanicals, including some foraged from Kew gardens itself.

The organic gin contains four categories of flavours; citrus (including lemon, lime and grapefruit peel); spice (nutmeg, liquorice and cassia); savoury notes (santolina, eucalyptus, rosemary and lemongrass); and florals (orris root and lavender flower). These botanicals are carefully infused for up to 24 hours, before the gin is distilled in a large copper alembic.

In 2019, Kew paired up with tonic water company East Imperial to add a 'Royal Botanic Tonic' to its range, using bark hand-selected from one of the last remaining cinchona plantations in Java. The bitter quinine extracted from the bark was paired with the sour notes of ruby red grapefruit and floral elderflower for a refreshing twist on the classic gin and tonic.

The founder of East Imperial, Kevin Law-Smith, recommends a gin and tonic made with 50 ml (1.7 fl oz) of Kew's original organic gin, 150 ml (5 fl oz) tonic water, a twist of grapefruit zest and a sprig of sage.

THE PRETTY BARMAID.

Pub^d by O. Hodgson, Cloth Fair, London.

PART IV
TONIC
COCKTAILS
"TABLOID"
Quinine
Bisulphate
BURROUGHS
WELLCOME & Co.
LONDON
"WELLCOME" BRAND
QUININE
SULPHATE
1 oz. (28.35 gm)
BURROUGHS
WELLCOME & Co.
LONDON
SYDNEY AND CAPE TOWN

Chère malade, c'est bien le véritable QUINA LAROCHE ;
il vous rendra la santé et vos belles couleurs.

8.

TEN WAYS WITH TONIC WATER

As we have seen, tonic water has a rich and complex history entwining stories of malaria, the cinchona tree, empire, and the cultural history of drinks.

The cinchona tree moved in various forms from the cloud forests of the South American Andes to early modern European apothecary jars and into the chilled glasses of British India. Few trees have had such an impact on the historypof the world. The popularity of its bark and alkaloids in tonic water in drinks, for health as well as pleasure, owes much to the refreshingly bitter, cooling properties of quinine, whether paired with gin or not.

Here we experiment with cocktails, with tonic water playing a central role. The chosen recipes reflect moments in cinchona's history that led to surprisingly drinkable combinations with spirits such as wine, brandy and rum.

And not forgetting quinine was valued by the temperance movement, we also give recommendations for tonic water to be enjoyed alone as the backbone to many delicious mocktails.

Cheers!

[opposite] Advertisement for Quina Laroche, a quinine-containing tonic wine (c. 1880). Wellcome Collection.

[previous page] The pretty barmaid (c. 1840). Wellcome Collection.

TALBOR'S 'ENGLISH REMEDY' TONIC

Robert Talbor's secret recipe (see p. 22) convinced two kings – the British Charles II and the French Louis XIV – of its effectiveness as a cure for malaria. After Talbor's death, it was revealed that his miraculous 'English Remedy' did indeed contain Peruvian bark, and its reputation was confirmed.

This recipe for a tonic gin is based on Talbor's famous 1682 formulation and includes his additions of rose petals, parsley and aniseed, but leaves out the cinchona bark and opium!

[opposite] Albarello drug jar used for cinchona bark, Spain (1731–70). Science Museum/ Wellcome Collection.

- 200 ml (6.7 fl oz) gin
- ½ teaspoon aniseed
- 1 tablespoon chopped fresh parsley
- 1 tablespoon dried chopped rose petals, or 2 tablespoons fresh

Method: Place all the ingredients in a 250 ml (8 oz) jar. Give it a shake to soak all the material and leave to infuse for 4–6 hours. Strain out the plant material and discard. Bottle the liquid.

To use: Use the infused gin as you would in the 'perfect gin and tonic' (recipe on p. 128). If stored in a cool dark place the mixture should last indefinitely.

CALCULATING QUININE

Caution! There are an abundance of recipes on the internet for making tonics and tonic syrups using cinchona bark but care must be taken when creating your own, as overdosage of the alkaloid quinine, which is a medical drug, can have unwanted side effects. Symptoms include flushed skin, ringing ears (tinnitus), impaired vision, abdominal cramps, dizziness, nausea and vomiting, among others. In addition, it should be noted that some people are allergic or hypersensitive to quinine. They should avoid it altogether, as should those who are pregnant, lactating or taking medications.

An instance of cinchonism was recounted by Clarissa Dickson-Wright, the television cook, in her autobiography. During a twelve-year period she drank four pints of tonic a day, to go with two pints of gin. Even once she gave up alcohol, the after-effects of the tonic gave her 'sticky blood'.

Cinchonism can be avoided by moderate use of home-made tonics and tonic syrups as well as careful calculation of maximum quinine content per millilitre. Tonic water brands use, and measure carefully, extracts of quinine to a maximum of 83 milligrams of quinine in a litre. Please use home-made tonic waters and syrups in moderation and with caution, discontinuing use if any adverse reactions occur.

Cort.
Cali=
say.

2. THE PERFECT GIN AND TONIC

For every gin you will find there is an ideal tonic for pairing – but provided you follow the basic gin-to-tonic ratio and use a decent quality version of both you won't go far wrong. So raise a glass in honour of the unknown hero who paired quinine, soda water and gin to create the classic cocktail.

- 50 ml (1.7 fl oz) gin
- 150 ml (5 fl oz) plain tonic water
- cubed ice (half fill the glass – a good cocktail does not skimp on ice)
- wedge of citrus (lime, lemon, orange or grapefruit)

Method: Though some argue the perfect glass is a balloon glass, a tumbler or high ball is fine. Place the ice in the glass, and carefully pour over the gin. Squeeze in and add your citrus wedge and top up with tonic water. Serve and enjoy.

3. THE ORANGERY

Kew's Triple Sec orange liqueur is inspired by its famous Orangery, built in 1761 for growing citrus fruits. It adds an orange twist to the classic gin and tonic, harking back to the orange quinine wines and orange peel that was used to dissolve quinine in early recipes.

- 25 ml (0.8–1.7 fl oz) gin
- 25 ml (0.7 fl oz) Triple Sec
- 150 ml (5 fl oz) plain tonic water
- dried orange slice or a twist of orange zest
- cubed ice

Method: Use a tall cocktail glass, half-fill with crushed ice. Pour on the gin and triple sec, then stir. Add the dried orange slice to garnish, then pour over the tonic water. Serve and enjoy.

Mocktail option: Tonic water with the juice of one orange and a twist of peel makes a delicious alcohol-free version.

[left] Juniper liqueur label illustrated with sprigs of juniper (19th century). Wellcome Collection.

[opposite] A Seville orange, *Citrus* x *aurantium* var. *melitense* by Pancrace Bessa. From F. Mordant De Launay and J. L. A. Loiseleur-Deslongchamps: *Herbier General de l'amateur* (1817–27).

P. Bessa pinx.
Barrois sculp.
Citrus Aurantium melitense.

Juniperus communis L.

GIN ST CLEMENT'S

Sparkling, quinine-containing Bitter Lemon was created by Schweppes in 1957 and remained on the pub-shelves ever since, used for the refreshing, non-alcoholic St Clement's cocktail. Orange juice is combined with lemon juice, or bitter lemon, and named after the bells of St Clement's, the first London church bell featured in the traditional nursery rhyme 'Oranges and lemons'. In this recipe we have added a gin twist, but feel free to leave out and enjoy a refreshing 'mocktail'.

- 50 ml (1.7 fl oz) gin
- 150 ml (5 fl oz) bottle of bitter lemon
- juice of a freshly squeezed orange (approx. 150 ml / 5 fl oz)
- cubed ice

Method: Place the ice in a tumbler. Pour over the gin followed by the bitter lemon. Finish with freshly squeezed orange juice to taste. Serve and enjoy.

Mocktail option: make without the gin for an alcohol-free version.

Botanical illustration of juniper (*Juniperus communis*), the berries are used to flavour gin. From *Köhler's Medizinal-Pflanzen* (1887).

GROG AND TONIC

One of the earliest sources for the use of alcohol to persuade soldiers to take their bitter quinine features rum, rather than gin. The combination of rum and tonic makes a surprisingly refreshing and delicious drink with a more earthy note than the simple gin and tonic. Based on the Navy Grog cocktail, which uses plain soda, this recipe adds a twist with tonic water.

- 50 ml (1.7 fl oz) dark rum
- 100 ml (3.4 fl oz) plain tonic water
- wedge of lime
- cubed ice

Method: Place the ice in a tumbler. Give the lime wedge a small squeeze and rub around the rim before adding it to the glass. Pour in the rum, followed by the tonic. Serve and enjoy.

6.

BRANDY TONIC

One of the earliest known tonic water cocktails – a 'ginger brandy and tonic water' – was drunk in Hong Kong in 1863, a mere five years after the patenting of tonic water. Though the source doesn't reveal its recipe, brandy works well with tonic, especially with a slice of ginger, to evoke this early cocktail. A delicate, sophisticated cognac like Hine or Delamain works best.

- 50 ml (1.7 fl oz) Brandy
- twist of orange peel
- slice of ginger
- plain tonic water
- cubed ice

Method: Place the ice, ginger and orange in a glass. Pour over the brandy, stir, then add the tonic water to taste. Serve and enjoy.

[opposite] Two huntsmen and a dandy smoking and drinking round a table (c. 1800). Wellcome Collection.

7.

PORT WINE AND TONIC

Known as porto tonico in Portugal, this refreshing mix is best served with a dry white port such as Taylor's Chip Dry White.

- 75 ml (2.5 fl oz) white port
- 75 ml (2.5 fl oz) plain tonic water
- cubed ice
- slice of lemon

Method: Place the ice and lemon in a glass and add the port and tonic. Serve and enjoy.

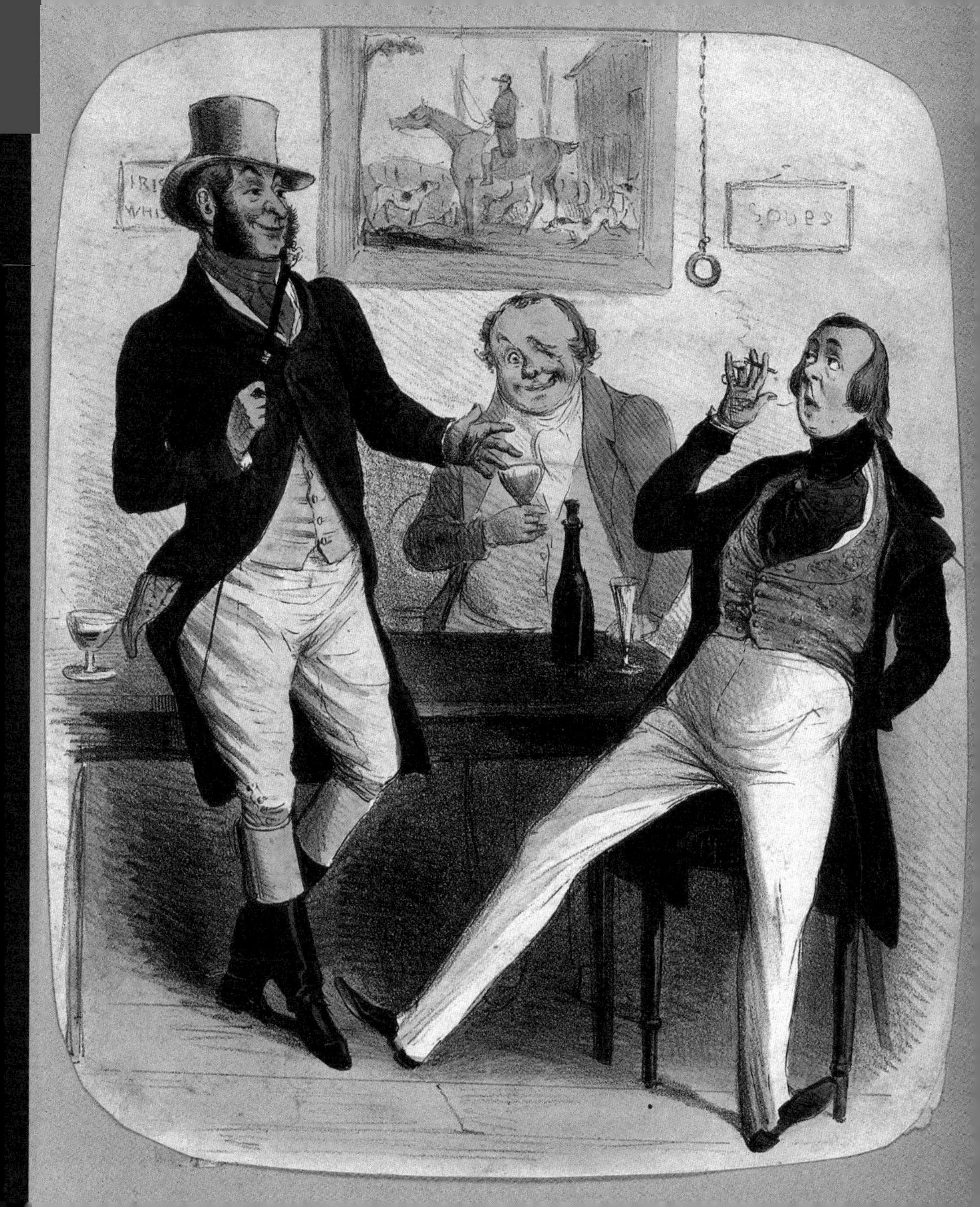

Vin de
Picardan
Habit de Cabaretier

8. TONIC SPRITZER

It might sound an odd pairing but think of how well lemonade and wine mix to create sangria. Evoking the early fever recipes combining ground cinchona bark with wine, this recipe makes a surprisingly refreshing drink. Light to medium-bodied, fruity red wines such as shiraz or tempranillo work well. If you prefer white wine, you could use a fruity, tropical New Zealand Sauvignon Blanc or an ebullient, spicy Alsace Gewurztraminer – but stick to a plain unflavoured tonic. Alternatively, try a fruity Viognier or Pinot Grigio with an added splash of delicious elderflower liqueur, such as St Germain.

- 150 ml (5 fl oz) red or white wine
- 150 ml (5 fl oz) plain tonic water

Method: Chill the wine and tonic water in the fridge, then mix together in a stemmed wine glass. Serve and enjoy.

9. WHISKY TONIC

Quinine whisky was promoted as a cure for colds in an American advertisement of 1895: '*it is pleasant to take, the bitter taste of quinine is disguised: It's a Success Wherever Introduced*'. A light, citrusy whisky such as the elegantly complex Glenmorangie, or Glenkinchie with its faint whispers of gingery spicy flavours, work well with tonic – but we cannot promise it will cure colds.

- 50 ml (1.7 fl oz) light whisky
- plain tonic water
- cubed ice

Method: Place the ice in a tumbler. Pour over the whisky and add tonic water to taste. Serve and enjoy.

[right] From *The Pacific Wine and Spirit Review*, San Francisco (1895).

[opposite] An innkeeper's suit of wine bottles and grapes (c. 1660). Wellcome Collection.

TONIC MOCKTAILS

CLASSIC TONIC MOCKTAIL

Tonic water was once a favourite of the temperance movement, as a mildly bitter, refreshing drink that combined health with pleasure and mixed well with fruity syrups. With more people preferring to go alcohol-free, it is time to re-embrace tonic water as the queen of the mocktail.

- 150 ml (5 fl oz) tonic water
- a long peel of grapefruit zest
- sprig of fresh thyme or rosemary
- cubed ice

Method: Place the ice in a tumbler. Lightly crush the sprig of thyme or rosemary, and the grapefruit zest before adding both to the glass. Pour the tonic water over slowly and stir before enjoying.

CREATE YOUR OWN MOCKTAIL

- fresh lemon, lime or orange juice
- crushed strawberries, a pinch of black pepper
- sprig of lavender
- ginger and rosemary
- rosemary and juniper berries
- rose petals and juniper berries
- elderflower cordial

THE G(RAPEFRUIT) & T

A mocktail version of the classic, this replaces gin with a few juniper berries and is finished off with a twist of grapefruit zest.

- 150 ml (5 fl oz) tonic water
- 5 juniper berries, lightly crushed
- a twist of grapefruit peel
- cubed ice

Method: Crush the botanicals lightly in a tumbler to release the flavours. Add the ice and tonic water, serve and enjoy.

Of all the Great Questions of these times, the most important is the health of man, and if you would be healthy use

THE QUININE TONIC CUP!

Health to all who will but seek it,
And when you have it, how to keep it;
This Cup will help you to regain it,
By constant use you may retain it.

This is one of Nature's own remedies which God has given to man for his use, and which for many years has been invaluable to thousands. It is in constant use in the UNITED STATES, and has been in this country for some years among the gentry, and highly valued by all who use it. It does not contain any of the mysterious dococtions or compounds placed there by chemical science, but it is as Nature gives it to us, and shaped by man for use.

'Tis from the bark of this wood that Quinine is made, and which is now so valuable to the whole world.

Fill this cup with water, let it stand a short time, the water becomes impregnated with Quinine and other medical properties which gives health to man.

DRINK IT MORNING AND NIGHT.

It will cleanse the blood from its impurities, and the skin from blotches by cleansing the stomach, aiding that function in its digestive powers, and give the stomach that tone which gives health, strength, and vigour to man.

'TIS HEALTH TO ALL WHO USE IT.

The Price is 6d.,

And no man will regret buying one, post free 7½d., from

THE TONIC STORES,

COTTON END, NORTHAMPTON.

[right] Handbill for the Quinine Tonic Cup (late 19th century). Wellcome Collection.

Reliable Quinine

For the treatment of malaria.
Of exceptional purity and alkaloidal value.

TRADE MARK **'TABLOID'** BRAND

Quinine Bisulphate

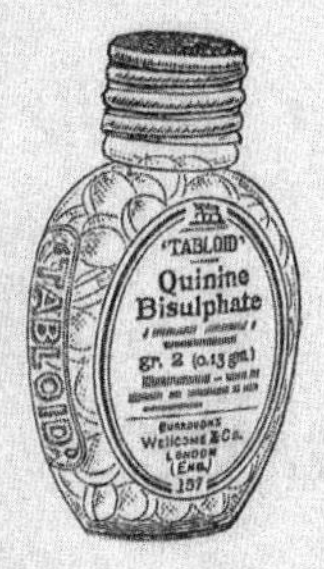

Famed throughout the world for its purity, accuracy, convenience and palatability.

Supplied as follows: gr. 1/2, in bottles of 50 and 100; gr. 1, in bottles of 36 and 100; gr. 2, gr. 3, gr. 4, gr. 5, gr. 10, 0·1 gm., 0·25 gm. and 0·5 gm., in bottles of 25 and 100. Issued *plain* or *sugar-coated*, except gr. 10 and 0·5 gm., which are *plain* only.

TRADE MARK **'WELLCOME'** BRAND

Quinine Sulphate

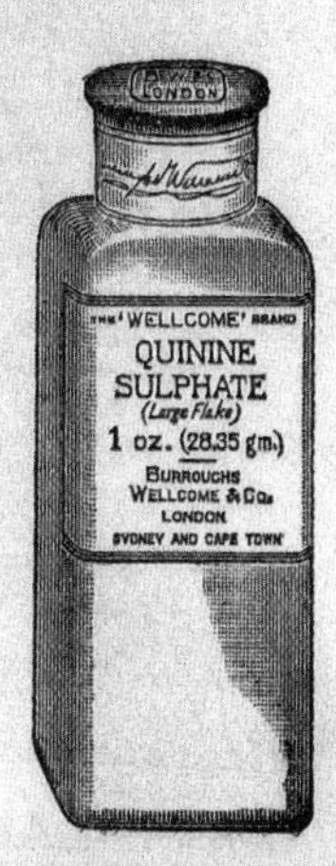

Attains a much higher standard of purity than required by the British Pharmacopœia.

Supplied in "Compact Crystals" and "Large Flake" (the ordinary form), both being identical in composition.

Issued in bottles and tins of convenient sizes.

Prices and supplies of all B. W. & Co.'s fine Quinine Products obtainable of Pharmacists in all parts of the world.

BURROUGHS WELLCOME & CO., LONDON
NEW YORK MONTREAL SYDNEY CAPE TOWN
MILAN SHANGHAI

XX 190 *Copyright*

Reliable quinine marketed by Burroughs Wellcome, c. early 20th century. Wellcome Collection.

FURTHER READING

1. THE MIRACULOUS FEVER TREE

Andersson, Lennart. 1998. A Revision of the Genus *Cinchona* (Rubiaceae–Cinchoneae). *Memoirs of the New York Botanical Garden* 80.

de Blegny, N. 1682. *The English Remedy, or, Talbor's Wonderful Secret for Cureing Agues and Feavers.* London: J. Wallis.

Bruce-Chwatt, Leonard J. *et al.* (eds). 1986. *Chemotherapy of Malaria.* Geneva: World Health Organization.

Crawford, Matthew J. 2016. *Andean Wonder Drug: Cinchona Bark and Imperial Science in the Spanish Atlantic, 1630–1800.* Pittsburgh: University of Pittsburgh Press.

Haggis, Alec W. 1941. Fundamental errors in the early history of *Cinchona. Bulletin of the History of Medicine* 10 (3): 417–59, (4): 568–92.

Honigsbaum, Mark, 2001. *The Fever Trail: the Hunt for the Cure for Malaria.* London: Macmillan.

Keeble, Thomas W. 1997. A cure for the ague: the contribution of Robert Talbor (1642–81). *Journal of the Royal Society of Medicine* 90: 285–90.

Maehle, Andreas H. 1999. *Drugs on Trial: Experimental Pharmacology and Therapeutic Innovation in the Eighteenth Century.* Amsterdam: Rodopi.

Pinault, Lauren L. and Fiona F. Hunter. 2012. Malaria in highlands of Ecuador since 1900. *Emerging Infectious Diseases* 18: 615–22.

Robertson, Robert. 1789. *Observations on Jail, Hospital, or Ship Fever.* New edition. London: The Author.

Rocco, Fiammetta. 2010. *The Miraculous Fever-Tree: Malaria, Medicine and the Cure that Changed the World.* London: Element.

Wallis, Patrick. 2012. Exotic drugs and English medicine: England's drug trade, c.1550–c.1800. *Social History of Medicine* 25: 20–46.

Webb, James L.A. 2008. *Humanity's Burden.* Cambridge: Cambridge University Press.

2. FEVERS AND AGUES

Chin, T. and Philip D. Welsby. 2004. Malaria in the UK: past, present, and future. *Postgraduate Medical Journal* 80: 663–66.

Cox, Francis. 2010. History of the discovery of the malaria parasites and their vectors *Parasites & Vectors* 3: 5.

Gachelin, Gabriel *et al.* 2017. Evaluating *Cinchona* bark and quinine for treating and preventing malaria. *Journal of the Royal Society of Medicine* 110 (1): 31–40, (2) 73–82.

Howell, Jessica. 2019. *Malaria and Victorian Fictions of Empire.* Cambridge: Cambridge University Press.

Hsu, Elisabeth. 2006. Reflections on the 'discovery' of the antimalarial qinghao. *British Journal of Clinical Pharmacology* 61: 666–70.

Institute of Medicine. 2004. *Saving Lives, Buying Time: Economics of Malaria Drugs in an Age of Resistance.* Washington, DC: National Academies Press.

Jarcho, Saul. 1993. *Quinine's Predecessor: Francesco Torti and the Early History of Cinchona.* Baltimore: Johns Hopkins University Press.

Karlen, Amo. 1996. *Man and Microbes: Disease and Plagues in History and Modern Times.* New York: Simon and Schuster.

Kuhn, Katrin G. et al. 2003. Malaria in Britain: past, present, and future. *Proceedings of the National Academy of Sciences* 100: 9997–10001.
Loy, Dorothy E. *et al.* 2017. Out of Africa: Origins and evolution of the human malaria parasites *Plasmodium falciparum* and *Plasmodium vivax*. *International Journal for Parasitology* 47: 87–97.
Marcus, Bernard A. and Hilary Babcock. 2009. *Deadly Diseases and Epidemics: Malaria*. Second edition. New York: Chelsea House.
Poinar Jr, George. 2010. Palaeoecological perspectives in Dominican amber. *Annales de la Société Entomologique de France* 46: 23–52.
Sallares, Robert. 2002. *Malaria and Rome: A History of Malaria in Ancient Italy*. Oxford: Oxford University Press.
Sherman, Irwin W. 2011. *Magic Bullets to Conquer Malaria: from Quinine to Qinghaosu*. Washington, DC: American Society for Microbiology.
Willcox, Merlin, Gerard Bodeker and Philippe Rasoanaivo. (eds). 2004. *Traditional Medicinal Plants and Malaria*. Boca Raton, FL: CRC Press.
World Malaria Report 2018. Geneva: World Health Organization.

3. PLANTATIONS AND POLITICS

Ainslie, Whitelaw. 1813. *Materia Medica of Hindoostan*. Madras: Government Press.
Brockway, Lucile H. 1979. *Science and Colonial Expansion: the Role of the British Royal Botanic Gardens*. New Haven, CT: Yale University Press.
Headrick, Daniel R. 1979. The tools of imperialism: Technology and the expansion of European colonial empires in the nineteenth century. *Journal of Modern History* 51: 231–63.
Howard, John E. 1862. *Illustrations of the Nueva Quinologia of Pavon*. London: Lovell Reeve.
Howard, John E. 1869. *The Quinology of the East Indian Plantations*. London: Lovell Reeve.
Keogh, Luke. 2019. The Wardian Case: Environmental histories of a box for moving plants. *Environment and History* 25: 219–44.
Markham, Clements R. 1862. *Travels in Peru and India while Superintending the Collection of Cinchona Plants and Seeds in South America, and their Introduction into India*. London: John Murray.
Roersch van der Hoogte, Arjo and Toine Pieters. 2014. Science in the service of colonial agro-industrialism: the case of cinchona cultivation in the Dutch and British East Indies, 1852–1900. *Studies in History and Philosophy of Biological and Biomedical Sciences* 47: 12–22.
Roersch van der Hoogte, Arjo and Toine Pieters. 2015. Science, industry and the colonial state: a shift from a German- to a Dutch-controlled cinchona and quinine cartel (1880–1920). *History and Technology* 31: 2–36.
Roy, Rohan Deb. 2017. *Malarial Subjects: Empire, Medicine and Nonhumans in British India, 1820–1909*. Cambridge: Cambridge University Press.
Veale, Lucy. 2010. An Historical Geography of the Nilgiri Cinchona Plantations, 1860–1900. PhD thesis, University of Nottingham.
de Vriese, Willem H. 1855. *De kina boom uit zuid Amerika overgebragt naar Java, onder de regering van Konig Willem III*. The Hague: C.W. Mieling.
Ward, Nathaniel B. 1852. *On the Growth of Plants in Closely Glazed Cases*. London: John Van Voorst.
Williams, Donovan. 1962. Clements Robert Markham and the introduction of the cinchona tree into British India, 1861. *Geographical Journal* 128: 431–42.

4. HUBBLE BUBBLE: THE HISTORY OF SODA WATER

Austerfield, Peter. 1983. Dr Phlogiston, the 'honest heretic'. *New Scientist* (24 March): 812–14.

Brownrigg, William. 1765. An experimental enquiry into the mineral elastic spirit, or air, contained in spa water; as well as into the mephitic qualities of this spirit. *Philosophical Transactions of the Royal Society* 55: 218–43.

Campbell, W.A. 1983. Joseph Priestley's soda water. *Endeavour* 7: 141–43.

Donovan, Tristan. 2013. *Fizz: How Soda Shook up the World*. Chicago: Chicago Review Press.

Emmins, Colin. 1991. *Soft Drinks: Their Origins and History*. Princes Risborough: Shire.

Götti, Robert P., Jörg Melzer and Reinhard Saller. 2014. An approach to the concept of tonic: suggested definitions and historical aspects. *Complementary Medicine Research* 21: 413–17.

Hughes, R. 1975. James Lind and the cure of scurvy: an experimental approach. *Medical History* 19: 342–51.

Kiple, Kenneth F. and Kriemhild C. Ornelas. 2000. *Cambridge World History of Food*. Cambridge: Cambridge University Press.

Petraccia, Luisa. 2006. Water, mineral waters and health. *Clinical Nutrition* 25: 377–85.

Priestley, Joseph. 1772. *Directions for Impregnating Water with Fixed Air: in Order to Communicate to it the Peculiar Spirit and Virtues of Pyrmont Water, and other Mineral Waters of a Similar Nature*. London: J. Johnson.

Riley, John J. 1958. *A History of the American Soft Drink Industry 1807–1957*. Washington, DC: American Bottlers of Carbonated Beverages.

Salzman, James. 2012. *Drinking Water: a History*. London: Duckworth.

Simmons, Douglas A. 1983. *Schweppes: the First 200 Years*. London: Springwood.

Talbot, Olive. 1974. The evolution of glass bottles for carbonated drinks. *Post-Medieval Archaeology* 8: 29–62.

van Tubergen, A. and S. van der Linden. 2002. A brief history of spa therapy. *Annals of the Rheumatic Diseases*. 61: 273–75.

5. EARLY MIXOLOGY: TONIC WINES, WATERS AND GIN

Ball, Ruth. 2017. *Rough Spirits and High Society: the Culture of Drink*. London: British Library.

Black, Rachel. (ed.) 2010. *Alcohol in Popular Culture*. Santa Barbara, CA: Greenwood.

Bowden-Dan, Jane. 2004. Diet, dirt and discipline: medical developments in Nelson's Navy. Dr John Snipe's contribution. *Mariner's Mirror* 90: 260–72.

Hands, Thora. 2018. *Drinking in Victorian and Edwardian Britain: beyond the Spectre of the Drunkard*. Palgrave McMillan.

Holt, Mack P. (ed.) 2006. *Alcohol: a Social and Cultural History*. Oxford: Berg.

Phillips, Rod. 2014. *Alcohol: a History*. Chapel Hill, NC: University of North Carolina Press.

Royal Botanic Gardens, Kew. 2015. *Kew's Teas, Tonics and Tipples*. Royal Botanic Gardens, Kew.

Stewart, Amy. 2013. *The Drunken Botanist*. Chapel Hill, NC: Algonquin.

Warner, Jessica. 2002. *Craze: Gin and Debauchery in an Age of Reason*. London: Profile.

6. ORIGINS OF THE GIN AND TONIC

Burnett, John. 1999. *Liquid Pleasures: a Social History of Drinks in Modern Britain*. London: Routledge.

David, Elizabeth. 2011. *Harvest of the Cold Months: the Social History of Ice and Ices*. London: Faber & Faber.

Henderson, James. 1863. *Shanghai Hygiene, Or, Hints for the Preservation of Health in China.* Shanghai: Presbyterian Mission Press.
Meyer, Christian G., Florian Marks, and Jurgen May. 2004. Editorial: Gin tonic revisited. *Tropical Medicine and International Health* 9: 1239–40.
Moore, William J. 1889. *A Manual of Family Medicine and Hygiene for India.* Fifth edition. London: Churchill.
Pack, James. 1983. *Nelson's Blood: the Story of Naval Rum.* Annapolis: Naval Institute Press.
Rees, Jonathan. 2013. *Refrigeration Nation: a History of Ice, Appliances, and Enterprise in America.* Baltimore: Johns Hopkins University Press.
Royal Commission on the Sanitary State of the Army in India. Vol II. Appendix 1863. London: HMSO.
Thomas, Jerry. 1862. *How to Mix Drinks, Or, the Bon-Vivant's Companion.* New York: Dick & Fitzgerald.

7. THE RISE, FALL AND RISE AGAIN OF THE GIN AND TONIC

Ashby, Maurice. 1942. British Empire drug production. *Journal of the Royal Society of Arts* 90: 138–51.
Barnett, Richard. 2011. *The Book of Gin.* New York: Grove.
Cheever, Susan. 2015. *Drinking in America: our Secret History.* New York: Twelve.
Ehmer, Kerstin and Beate Hindermann. 2015. *The School of Sophisticated Drinking.* Vancouver: Greystone.
English, Camper. 2016. *Tonic Water AKA G&T WTF.* Kindle e-book.
Estes, J. Worth. 1988. The pharmacology of nineteenth-century patent medicines. *Pharmacy in History* 30: 3–18.
Gately, Iain. 2011. *Drink: a Cultural History of Alcohol.* New York: Gotham.
Helstosky, Carol. (ed.) 2014. *The Routledge History of Food.* Abingdon: Routledge.
Jeffreys, Henry. 2016. *Empire of Booze.* London: Unbound.
Nicholls, James. 2009. *The Politics of Alcohol: a History of the Drink Question in England.* Manchester: Manchester University Press.
Ocejo, Richard E. 2010. What'll it be? Cocktail bartenders and the redefinition of service in the creative economy. *City, Culture and Society* 1: 179–84.
Office for National Statistics. 2018. *Adult Drinking Habits in Great Britain: 2017.* Newport: ONS.
Smith, Andrew F. 2014. *Drinking History: Fifteen Turning Points in the Making of American Beverages.* New York: Columbia University Press.
Solmonson, Lesley J. 2012. *Gin: a Global History.* London: Reaktion.

Quinquina Dubonnet, by Jules Chéret (1895). Victoria and Albert Museum, London.

INDEX

Images are indicated by page numbers in **bold**.

ABOUT THE AUTHORS

KIM WALKER trained as a medical herbalist, and now specialises in the history of plant medicines. She is currently working on a PhD on cinchona at the Royal Botanic Gardens, Kew and Royal Holloway, University of London. She is on the committee of the Herbal History Research Network, the British Society for the History of Pharmacy and is a member of the Association of Foragers. She is the co-author of *The Handmade Apothecary* (Kyle Books, 2017) and *The Herbal Remedy Handbook* (Kyle Books, 2019).

MARK NESBITT is curator of the Economic Botany Collection at the Royal Botanic Gardens, Kew. His research centres on botany and empire in the nineteenth century, and on the history and current-day relevance of botanical collections. He is the co-author of *Curating Biocultural Collections* (Kew Publishing, 2014) and *The Botanical Treasury* (André Deutsch, 2016). Mark is a Visiting Professor in the Department of Geography, Royal Holloway, University of London.

PICTURE CREDITS

All images unless otherwise stated © The Board of Trustees of the Royal Botanic Gardens, Kew.

We are very grateful to the following for use of images:

© ABCR Auctions: 80 (top right)

Allard Pierson, University of Amsterdam [077.580]: cover (plant illustration), 9

British Library: 64, 71 (left), 112

© British Library Board. All Rights Reserved / Bridgeman Images: 82

East Imperial: 120 (right)

evgdemidova: cover (mosquito)

© Florilegius / Alamy Stock Photo: 8

© Imperial War Museum: 116–7 (FRE 10686)

Library of Congress: 104–5 (all)

Metropolitan Museum of Art: 12–13

National Library of Medicine: 27, 93

New York Public Library: 114, 119

Rijksmuseum: 110

Science Museum/Wellcome Collection (Creative Commons CC BY, creativecommons.org): 69, 127

Smithsonian Libraries: 72–3

State Library Victoria: 80 (top left)

University of Minnesota Libraries: 99

© Victoria and Albert Museum, London: 141

Wellcome Collection (Creative Commons CC BY, creativecommons.org): 7, 16, 17, 20, 21, 22, 24, 26 (all), 28–9, 30, 31, 32, 34, 35 (all), 36 (all), 37, 39, 49 (left and right), 58 (top and middle), 59 (all), 60, 62, 65, 66–7, 68, 70, 71 (right), 75 (top left and right), 76, 78, 81, 83, 86, 87, 88, 92, 95, 96-7, 98, 100, 105 (all), 108, 111, 122, 124, 128, 132, 133, 134, 136, 137

Wikimedia Commons: 33, 38, 74

Every effort has been made to acknowledge correctly and contact the source and / or copyright holder of each picture and the Royal Botanic Gardens, Kew apologies for any unintentional errors or omissions which will be corrected in future editions of this book. Please contact publishing@kew.org.

An immaculately researched, beautifully written, gorgeously illustrated history of tonic water in all its forms. We wouldn't be drinking gin in such vast quantities today if it wasn't for the tonic we slosh into it and this delightful romp through the beverage's history tells you everything you need to know about this vital panacea. The perfect book to dip into as you swat that bloody mosquito and ponder your first G&T of the day.

Jonathan Ray, author and drinks editor for *The Spectator*

Discoveries from this latest fact-finding expedition reveal nearly everything I learned about tonic water is a myth. The true story told here traverses the globe; from the age of exploration through the Industrial Revolution and beyond, before dropping readers off in the midst of a mixed drink renaissance. From fever trees to pharmacies and mixology; few tipples team with a tale as beguiling and quixotic as quinine.

Jim Meehan, author of *The PDT Cocktail Book* and *Meehan's Bartender Manual*

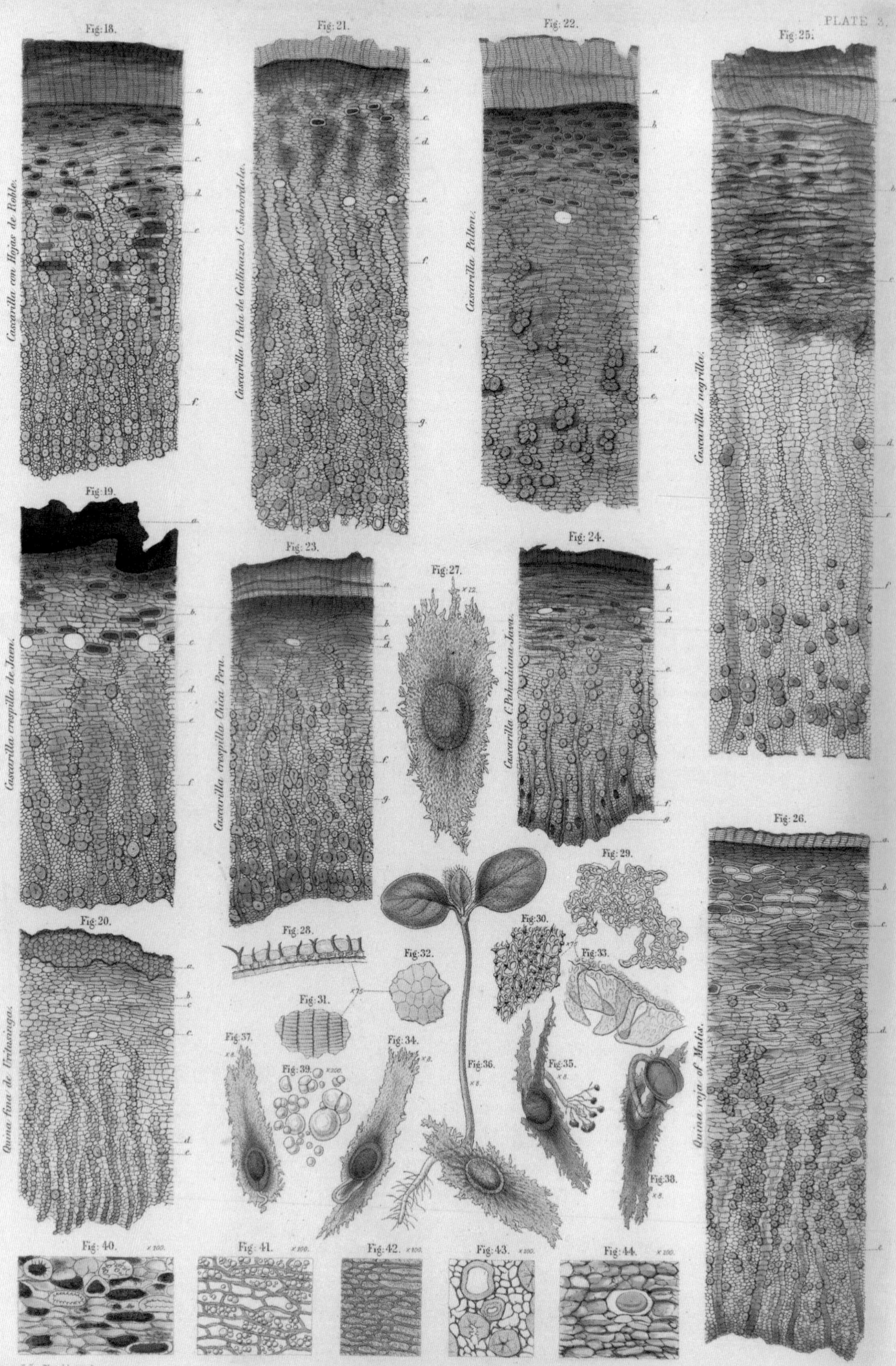

All magnified 50 diameters, except where otherwise marked.